SPACE PIRACY

SPACE PIRACY

Preparing for a Criminal Crisis in Orbit

Marc Feldman and Hugh Taylor

WILEY

Published by John Wiley & Sons, Inc., Hoboken, New Jersey.
Published simultaneously in Canada and the United Kingdom.

ISBNs: 9781394240203 (paperback), 9781394240227 (ePDF), 9781394240210 (ePub)

For general information on our other products and services, please contact our Customer Care Department within the United States at (800) 762-2974, outside the United States at (317) 572-3993. For product technical support, you can find answers to frequently asked questions or reach us via live chat at https://support.wiley.com.

If you believe you've found a mistake in this book, please bring it to our attention by emailing our Reader Support team at wileysupport@wiley.com with the subject line "Possible Book Errata Submission."

Wiley also publishes its books in a variety of electronic formats. Some content that appears in print may not be available in electronic formats. For more information about Wiley products, visit our web site at www.wiley.com.

Library of Congress Cataloging in Publication data available on request.

Cover images: Helmet: © Jefry Maviskho/Adobe Stock,
             Crossbones: © martialred/Adobe Stock
Cover design: Wiley

SKY10093893_122024

For Leslie and Rachel

Contents

Foreword

I've spent the better part of three decades serving in the United States Air Force, and now the Space Force, working to deter conflict and ensure that the United States leads the world in space. The capabilities I have contributed to developing include space communication, space control, missile warning, intelligence/surveillance/reconnaissance, position/navigation/timing, weather, and access/mobility/logistics—the whole gamut. Success in these critical missions requires that we perform constant assessments of space-related threats.

For two decades following the Cold War, our threat analyses focused almost exclusively on perils naturally occurring in the space environment. Following the destructive Chinese antisatellite missile test in 2007, we pivoted to primarily focus on the threats from China and Russia, potential adversaries powerful enough to field their own space programs and weapons. Today the list of potential nation-state threats has grown, but we are still focused on deterring only sovereign states. This is about to change, in my view, and this book offers a glimpse of what is coming: irregular threats to space, especially crime and piracy.

The authors argue persuasively that space piracy is inevitable. As the costs and technical barriers to accessing space continue to come down, new wealth will be generated in space. At least some of that wealth will be able to be stolen or held hostage. Many state and nonstate actors will be capable of space piracy; some will inevitably act upon the opportunities they see.

As for the timing of this threat, we don't know when space piracy will become a real issue, but it will arrive at some point, perhaps sooner than we imagine. Some forms of space piracy are possible today, such as radiofrequency jamming and cyber hostage-taking. More stereotypical piracy, such as stealing precious metals mined from asteroids, may or may not be decades away.

Regardless, with such an uncertain timeline, now is the time to consider the threat and develop our strategies, tasks that require few resources. If developed and executed well, our strategy could very positively shape our future in space.

What should we do about the threat of space crime and piracy? In my view, we can deter and defeat space pirates, but not by doing nothing. The book thoughtfully outlines specific actions that can and should be taken today, specific actions for the intelligence community, Space Force, private sector, and other stakeholders. We need to build upon the existing international frameworks and institutions, such as the Outer Space Treaty and United Nations, and learn from the history of tackling piracy in other domains.

This is a thought-provoking book written for everyone who builds or relies upon space capabilities—and that is almost everyone. Every Space Force Guardian needs to read it so that we can contemplate and debate these critical topics. Just as the Navy counters piracy on the high seas, the Space Force is the United States' first and best choice for mitigating the threat of space piracy. To this end, the topic of space piracy needs more intellectual attention, discussion, and strategizing. Space Piracy is a tremendous first step in that direction.

—Col. Eric Felt, US Space Force
Director, Space Systems Engineering
Office of the Assistant Secretary of the Air Force
for Space Acquisition and Integration

Preface

When Neil Armstrong exited the lunar module on July 20, 1969, and became the first human being to set foot on the moon, he famously said, "That's one small step for man, one giant leap for mankind." It was one of the most awesome and inspiring moments in history. Through remarkable feats of science and engineering, a man had ventured forth into outer space "for mankind," coming in peace as a representative of the noble aspirations and capabilities of the entire human race.

Armstrong's sentiment also appeared in the groundbreaking Outer Space Treaty in 1967. It alludes to "the common interest of all mankind in the progress of the exploration and use of outer space for peaceful purposes."

Today, after half a century of engineering breakthroughs, outer space is rapidly transforming into a commercial resource that is available for the use of all humankind. And, for better or worse, the broadening exploitation of space is exposing the ignoble side of humanity, with the potential for crime, corruption, piracy, and war in space. Indeed, this is already happening, though at a small scale. It's early in space's transition from noble representation of the best of humanity to a spectacle of its ignobility, but the moment has arrived, and we ignore it at our peril.

In this regard, science fiction has been well ahead of practical realities, depicting space as a place where anything goes, and anything will. George Lucas saw it coming in 1977, with the notorious bar scene in *Star Wars* giving us a preview of the dark underbelly of outer space, though *Star Trek* was already trafficking in space criminality a decade before. Reality is starting to catch up with science fiction, though, more quickly than many of us might imagine. As space becomes the site of wealth creation, a process that is already underway, it will almost certainly become the site of criminal activities, including acts of piracy.

Are we ready for piracy and crime in space? This is the theme we intend to explore in this book. For some, the idea of bad behavior in space is anathema, an affront to their sense that space should be a zone of human cooperation and heroic endeavor. For others, it has long been clear that space has the potential to become a lawless cesspool featuring the worst that humanity has to offer.

The time has come to reconcile these conflicting views. We will try, suggesting that the best of humanity's institutions, from the law to intelligence services, corporations, the sciences, and the military, can come together to combat what the worst of us may be dreaming up.

Introduction

In 1929, a 17-year-old German boy named Wernher Magnus Maximilian Freiherr von Braun bought a movie ticket that would change his life and, eventually, the world (see Figure I.1). In a Berlin cinema, he watched what is considered the first great science fiction epic: Fritz Lang's two-and-a-half-hour silent film *Frau im Mond* (Woman in the Moon).

FIGURE I.1 Dr. Wernher von Braun at the Marshall Space Flight Center in Huntsville, Alabama, 1960.
Source: NASA

The movie tells the story of a visionary scientist, Professor Mannfeldt, who has a plan to send a rocket to the moon, where he believes there is a fortune in gold. A gang of thieves steals the professor's plans and threatens to destroy his mission if he doesn't cut the gang in on it. He agrees and sends his assistant, a woman named Friede, to the lunar surface, where she discovers breathable air and lots of gold. But the whole situation goes sideways, and the thieves try to hijack the rocket, destroying it in the process.

The depictions of the rocket inspired von Braun to devote his life to rocket science. In fact, they were so realistic that the Gestapo banned the film after von Braun commenced his top-secret work on the notorious V2 rocket, which devastated London in World War II. After the war, von Braun enabled the Americans to put a man on the moon with the Apollo missions— just 40 years after he watched Lang's film (see Figure I.2).

Even as space exploration was in its infancy, forward-looking people recognized that space could be a venue for piracy, a form of criminal mischief and war by other means that has existed for millennia. A 1952 *Colliers Magazine* article asked, "Who Owns the Universe?" and anticipated the threat of pirates operating in space. We're going to start answering that question in this book.

The current security conditions in space are ideal for piracy. Space systems are vulnerable. Great riches await pirates who can breach their poor defenses. The talent pool of available pirates is growing as nations with space expertise become economically and geopolitically unstable. Global criminal cartels allied with rogue state actors are poised to exploit these conditions for profit.

It is our contention that as the world embarks on lucrative ventures in space, it is a virtual certainty that space piracy will soon emerge as a major threat to their success. This may seem hard to believe, but the world has changed since the heady days of the Apollo program. The conditions that favor space piracy are on the rise.

While space exploration has long been the province of national governments, an honorable form of national service and international cooperation, space is rapidly transforming into a corporate free-for-all and playground for rogue states. Stakeholders will ignore these dramatic shifts in policy and ethics at their own peril.

FIGURE I.2 Still frame from Fritz Lang's 1929 silent film *Frau im Mond* (Woman in the Moon), juxtaposed with a photo of the German V2 rocket, shown at the Rocket Center of the Third Reich in Peenemuende, Germany and the poster for Fritz Lang's 1929 silent film "Frau im Mond" (Woman on the Moon)"
Sources: Nataliia Budianska/Shutterstock and Pictorial Press Ltd/ Alamy Images

Today, nearly a century after *Frau im Mond,* space is home to a large, growing collection of industries. Corporations are following in the footsteps of Professor Mannfeldt. They are contemplating mining for gold on asteroids like modern-day incarnations of the professor. And, as the plot of the film suggested even in 1929, space was a rich venue for criminal conspiracies. *Frau im Mond* was disturbingly prescient.

It is our view that piracy in space will become a lethal, vexing, and technologically sophisticated aspect of the new space economy. It will force all the stakeholders in space to rethink how they operate in space. It will also lead to a reckoning as to the efficacy of legal, military, and global enforcement bodies in dealing with crime, piracy, and smuggling in space.

With all of this in mind, let's begin.

Can You Actually Predict the Future?

Before we dig into the topic of space piracy, we feel a need to address a question that has come up multiple times in the process of writing this book. That is, "Who are you to imagine that you can predict the future?" It's been asked more politely than that, but the gist is the same. This book anticipates events that have not yet occurred, though there have been some incidents that show clear signs of what's coming. Who are we to prognosticate?

This is indeed a problem. Serious people don't consult crystal balls and expect to be taken seriously. Only a charlatan would predict the future, right? The answer is yes and no. Yes, making guesses about what is going to happen next year, or in 20 years, requires an admission, up front, that you could be wrong, and probably are. And yes, only a charlatan would make a claim like, "Tomorrow, the Dow Jones will go up 102 points."

At the same time, many of us, including a lot of highly paid and very serious people will say things like, "If the Federal Reserve raises interest rates next week, there is a high probability that the Dow Jones will go down." That is a prediction of a future event that is based on past experience. It is not absolute. Rather, it is framed in terms of probability. In this sense, you can predict the future.

We feel comfortable predicting that piracy and other forms of mass criminality will arise in space as the sector grows more commercialized. Like the stock analyst who expects the Dow to drop if interest rates go up, we too anticipate that criminals will target rich space assets based on past experience. In our view, just as pirates over millennia have tried to seize ships carrying valuable cargoes when they transited narrow sea passages, so too will space pirates attempt to hijack space riches when they have the opportunity. This is not a fanciful guess. It's a prediction with a high probability of coming true.

Furthermore, if we take a step back from this immediate topic, we can see that predicting the future is the foundation of much policy planning.

The Pentagon, for instance, strategizes, recruits, and spends based on future warfighting scenarios. Does the Pentagon know for sure that Country A is going to attack Country B, drawing the United States into the conflict? Of course not, but they have to plan for that scenario. We have military units formed and ready to address conflicts that have only a probability of occurring.

Whom This Book Is For

Who should read this book? In our writing process, we have spoken with many truly impressive people who have much more specific knowledge in their areas of specialty than we do. Our intention is not to replicate their entire expertise here. Indeed, each of these people could write their own books, and several have. Instead, our approach is to be integrative. The book blends our own knowledge with that of our subject-matter experts.

This book is for anyone who wants to gain a broader understanding of the potential for space piracy and crime than they already possess. If you're from the finance field, for example, you will gain insights into the geopolitical aspects of space piracy from this book. If you're from the world of intelligence, you may learn a thing or two about space insurance, and so forth. Our goal is to broaden your horizons with a (we hope) insightful overview of the subject.

Not Authoritative, by Design

This book is not authoritative in nature. That's on purpose. For one thing, it's hard to be authoritative on a topic so speculative. More importantly, dealing with space piracy is by necessity a process that has to involve experts from a range of disciplines. There are no individuals who can be authoritative about space piracy. The subject is too multithreaded, too integrative for that, at least in the present. As time goes on, we may see space piracy experts, people who can speak with authority on the topic. Today, however, our goal is to introduce the full breath of subjects that come under the rubric of fighting space piracy and initiate the interdisciplinary dialogues that will ideally make progress on this serious issue.

Space Piracy, an Overview

What do we mean by *space piracy*? A simple answer is robbery or hijacking of objects of value in space, akin to piracy on the high seas. Just as pirates board oil tankers off the coast of Africa and hold them for ransom, we envision space-based pirates hijacking space cargoes and demanding payment to set them free.

The specifics of how space piracy would work remain to be seen, but in all probability, space piracy will be significantly more complex than piracy at sea. For one thing, space piracy requires the use of costly, sophisticated space equipment. The barriers to entry are higher in space than they are on the world's oceans.

Space piracy will also probably involve the use of robots or other automated technologies to perform many of the acts of piracy. Human space travel is rare and very expensive. Though it may be entertaining to imagine a Johnny Depp character swashbuckling on the lunar surface, the greater likelihood is that someone on Earth will direct a robot to hijack the item of value and hold it for ransom.

Additionally, space piracy may encompass criminal acts that take place on Earth. This might mean hijacking launch stations and satellite uplink stations. In fact, in our view, space piracy could take place across at least four different kinds of space, including outer space, cyberspace, terrestrial space, and virtual/AI-generated space.

A useful definition of space piracy might also include a broader range of criminal activities, analogous to those that take place at sea today. For example, smuggling, which is not piracy per se, falls into a comparable zone of law enforcement (or lack thereof). Similarly, trafficking in sanctioned goods, a la today's Russian "Dark Fleet" oil tankers, could have a corollary in space. Such acts trigger the same questions of space sovereignty, space law, and space policy as space piracy. For this reason, it makes sense to include them in our analysis.

Is This Book Necessary?

The other question we've been asked a few times is "Why are you writing this? The problem doesn't exist yet." That's a fair question. The world is experiencing plenty of real problems in the current moment. Why worry about something that hasn't happened?

We should be worrying about space piracy right now. For a reference point, take a hard look at the state of maritime piracy. You will see a criminal problem that lacks effective solutions. There are myriad reasons for this, but the most serious issues have to do with a lack of a coherent body of law or policy to rein in piracy on the high seas. Many of the same problems exist in space.

Without action, space pirates will have an open field for the most outlandish and costly criminal activities—if we don't prepare to combat them. The ongoing disruptions by nonstate actors like Yemen's Houthis, who are wreaking havoc on world shipping, further underscore the problems that come from being unprepared to deal with irregular and asymmetric adversaries.

The other question that comes up is "Why now?" Space commerce is in its infancy. The kind of high-value space assets and industries we contemplate in this book are decades, if not centuries, in the future. Why deal with this issue now? Isn't there plenty of time to prepare?

Yes and no. We needn't rush. Figuring this out could take a few years, but now is actually the ideal time to start preparing for this coming criminal crisis in space. We have several reasons for taking this stance. For one thing, bad policies tend to harden and become difficult to change. We see this with maritime piracy, where centuries of maritime law have created a virtual straitjacket for shippers and law enforcement agencies that want to fight piracy. If we want to avoid such an outcome in space, we need to start figuring out better policies today.

Cybersecurity offers another cautionary tale. The world is in the midst of a massive cybercrime wave that costs at least $200 billion a year to deal with, and the results are pretty abysmal. Computer systems remain insecure, despite the fortunes being spent to protect them. The current cyber disaster is the direct result of bad decisions made about computer and network security 60 years ago. At the time, few people envisioned a global network connected to billions of personal devices.

It would have been good if they had, because they could have designed more security into the standards and products we use today. But, they didn't, and our data is vulnerable to breach every day. With this lesson in mind, we should look at space technology and policy with a similar eye to futuristic crime. What can we design more securely today to avoid a criminal catastrophe in 50 years?

Another question that we've encountered is "Does it matter?" The basis for this question is that piracy is a fact of life on the high seas, but life goes on. The world's shipping lanes and supply chains all function, with some allowance for losses due to piracy. Space industry, too, would probably survive piracy, so why bother with it? It's a good question, but in our view, a better question to ask would be "If we can avoid a major problem, why hesitate?" Why ignore a threat that will surely have an impact on space commerce, a threat that could mask hostile action by geopolitical adversaries? The technology used by space pirates will probably be more lethal than that used by today's sea pirates. The problem is at least worth a serious look in the present day.

If the time horizon bothers you, consider the Treaty of Versailles, which ended World War I. It's been 104 years since the treaty went into effect, and the world is still convulsing from its impact. The endless wars in the Middle East are but one example of how a 104-year-old document set massive, intractable conflicts into motion.

The decisions we make about space policy today will have ramifications for hundreds of years. It's on us to gather our wisdom and develop policies, legal frameworks, and military/intelligence organizations that can keep space safe from crime and piracy. It's a big responsibility, but we think we're up for it.

What's Inside

Our goal in this book is to introduce the idea of space piracy, describe how we think the phenomenon will occur, and then offer some suggestions for mitigating the risk. We refer to our content as "speculative nonfiction." We start by sharing some potential space piracy scenarios in Chapter 1, which we have written in a second-person "this is happening to you" voice to make the risks and impacts of space piracy vivid to you, the reader. We cover possible space piracy scenarios that could take place today, as well as in the future. Chapter 1 concludes with some thinking about the likelihood of these scenarios becoming realities.

In Chapter 2, we digress into a brief history of piracy, starting with piracy in the ancient Mediterranean Sea and continuing up through the "golden age" of piracy in the seventeenth and eighteenth centuries. We define and discuss the respective stories of privateers, who were given "letters of marque" by their monarchs to plunder enemy vessels, and buccaneers, who simply stole to get rich.

We go into some detail on the history of piracy because we believe the past is prologue. The same kinds of forces and power vacuums that led to widespread piracy in the past are looming in the future of space. The old piracy provides a great platform to look at piracy as a tool of geopolitics, and several other important lessons about how the world actually works, versus how we want or think it works.

We continue from there in Chapter 3 to an analysis of the emerging multitrillion-dollar space economy. This is a necessary part of the story, because if there isn't going to be any wealth in space, there will be nothing to plunder. We look at the near-term market for satellite industries and related fields like in-orbit servicing. We explore the potential for lunar industries, including lunar mining and the use of the moon as a platform for launching large-scale space missions of all types.

After laying out the tradition of piracy and discussing how it is endemic to the human condition, Chapter 4 gets down to defining space piracy in more detail. We offer potential structures of space pirate organizations and how they might operate. This includes a look at some of the leading criminal organizations, such as the major crime cartels. These groups have the money and organizational cohesion to pull off crimes in orbit. This chapter also differentiates space privacy from space warfare. It considers the potential for privateering in space.

Chapter 5 devotes itself to space hacking. We felt this subject deserved a significant degree of attention for several reasons. For one thing, space hacking is occurring today, and it is arguably the first budding of what will become a much larger space pirate enterprise as the future unfolds. Additionally, given that space vehicles are platforms for computer technology, their vulnerabilities are worth exploring as a way to get a better sense of how space commerce can be disrupted by piracy.

Chapter 6 takes on space law, policies, and treaties. This is a mammoth topic that is central to the entire discussion of space piracy and crime.

Entire books have been written on this, so we're just scratching the surface. Nevertheless, we endeavor to give you as full an understanding as possible of the existing legal frameworks in space, as well as their limitations. Space is a "commons," similar to the world's oceans. For this reason, as we explain, maritime law provides a roadmap for what law in space may look like as it evolves.

It's not a simple or soothing subject. There are a lot of dangerous blank spots in the space law landscape. Sovereignty is muddled. Enforcement is effectively impossible, even when the law is clear, and that's not very often. The maritime practice of "flags of convenience" will likely surface in space, further complicating efforts to define and enforce laws within space's meager legal framework.

Having discussed the underlying issues defining space piracy and law, the book then turns to its more specific potential impacts. Chapter 7 goes into greater detail answering a question posed in Chapter 4, which is "Who will be the space pirates?" We don't know, of course, but we will make some educated guesses. In our estimation, the best candidates for space piracy are the existing criminal cartels. They have the resources and strategic resolve to enter this expensive but lucrative field of crime.

Chapter 8 deals with the potential for space piracy and crime to affect national security. We will focus on the United States, but space is a factor in national security for every nation. We will also challenge some prevailing assumptions, such as that China looms as the aggressor in this context. This is not necessarily true, and it's valuable, in our opinion, to question the existing thinking.

We will explore the interplay between three interlocking realms that affect national security in space. They are sovereignty, criminal organizations, and private enterprises. The latter category may include corporations that either have or arrogate the ability to act as quasi-sovereigns in space like the British East India Company did in the eighteenth and nineteenth centuries.

Chapter 9 looks at space piracy and crime from the perspectives of business and finance. We will examine the potential for money laundering in space. We will also discuss the potential for legitimate businesses to become inadvertent vehicles for space piracy, such as through the work of corrupt insiders—but also the possibility that criminals will take over legitimate businesses and use them for space piracy purposes.

In Chapter 10, we answer the big question that should be on your mind as you read this book: "What can be done to prevent space piracy and crime from becoming a huge problem?" It is going to be impossible to stop all space crime, but we now have a chance to think through ways to mitigate the worst of it. Chapter 10 offers policy recommendations and counter-measures to reduce the impact of space piracy and crime. These include some legal recommendations, ideas for new space treaties, and suggestions for secure engineering of space technology.

Chapter 10 then takes policy ideas into the realm of practicality. Who will enforce these policies? As at least one former member of the intelligence community mused, "That is a really good question...." We believe that combatting space piracy and crime will only partly be a matter for law enforcement. The transnational and questionable sovereignty issues in space will disempower traditional law enforcement in space.

Rather, we think that intelligence services, along with the armed forces, will be front and center in dealing with space piracy. What this will actually look like remains to be seen, but the good and bad experiences in confronting maritime piracy offer some clues as to how this might be best be handled. We will also contemplate some novel organizational constructs for intelligence and alliances between previously disconnected entities that will serve the mission of fighting space piracy and crime.

Possible steps include empowering the U.S. Space Force to fight space pirates. Shoring up space cyber defenses and updating international treaties will ensure the United States has unfettered freedom to combat space piracy in space and on land. Going further, we discuss the development of novel public/private cooperation between intelligence agencies, private firms, law enforcement agencies, and others.

About Us

As coauthors, we come to this subject from different backgrounds. Marc Feldman has been active in the business of space since the 1980s, when he was instrumental in setting up Televisa's first satellite operations. Since that time, he has advised and invested in a variety of space ventures. He has also worked in the intelligence field. Hugh Taylor is a professional writer and author who has worked in the fields of enterprise technology and cybersecurity for over 20 years.

As we set about writing the book, we knew that we lacked deep knowledge in many critical subject areas. As a result, we sought the best people we could find in fields like space law, geopolitical strategy, and so forth. The book is intended to read, in part, as a series of conversations with top thought leaders who have unique perspectives and insights into their respective subject areas.

How to Read This Book

You can read this book front to back, and we hope you do. However, if you are already knowledgeable in one of our focus areas, you can read chapters selectively and come away—we hope—better informed. Our goal is to offer a stimulating dialogue, rather than attempt to be expert in all areas. Indeed, in our view, true expertise does not exist for this subject. In that context, we suggest reading the book in any way that adds to your understanding of the issues.

1 A Not-So-Unlikely Scenario Coming to the Space Near You

Depiction of communication networks that link through space
Source: NicoElNino/Adobe Stock Photos

When we were kids, that distinctive voice of God that narrated every movie trailer used to thunder "Coming soon…to a theater near you…" before proceeding to thrill us with the latest coming attractions. As the worlds of space exploration and space industry continue to grow and evolve, it's useful to imagine some coming attractions, if you will. Not fun, exciting coming attractions. Rather, the stakeholders in the space sector would be wise to consider risks they may be facing in the near—and longer—term future. And, of course, some attacks are already taking place today.

The Value of Scenario Planning

We have backgrounds that include cybersecurity and intelligence. In both fields, scenario planning can be an effective way to model risk and think through issues of prevention, detection, and response. It also brings jurisdictional issues into sharp relief. The scenario building process lets you use your imagination to come up with possible avenues of attack and defense that may not come to mind if you're just looking at risk in an objective, academic sense.

Some Scenarios to Ponder

With that in mind, we will share a number of possible space piracy scenarios. Some are likely. Others are more far-fetched, though even those have instructional value. Some are presently occurring, while others are off in a future that's arriving a lot sooner than many of us want to admit.

Present-Day Risks: Pirates and Cyberattacks on Satellites

Space pirates tend to be highly ad hoc, improvisational groups that seek the easiest targets for plunder. In the space economy, the easiest and most accessible targets are the more than 8,000 satellites in orbit,[1] a number that is expected to double in the next five years. These satellites, including commercial and military, provide various services to their customers, from commercial broadcasting and imaging/observation to missile tracking and shipping.

Satellite communications (satcom) plays a crucial role in many government, military, and commercial sectors. Pirates know they can potentially control satcom. The ability to disrupt, inspect, modify, or reroute traffic provides a valuable opportunity to perform surveillance or conduct cyberattacks. Terrestrial network infrastructures are subject to physical limitations and simply are unable to meet demand.

To fill this gap and provide improved performance, there are multiple satellite constellations orbiting Earth. These networks are responsible for, among other things, allowing people in remote locations to access the Internet, helping vessels and aircraft operate safely, and providing military and emergency services with critical communication links during armed conflicts or natural disasters. Sectors that commonly rely on satellite networks include aerospace, maritime, military and governments, emergency services, telecommunications, industry (oil rigs, gas, electricity), and the media.

Satcom infrastructure can be divided into major segments: space and ground. Space includes those elements needed to deploy, maintain, track, and control a satellite. Ground includes the infrastructure required to access a satellite repeater from Earth station terminals. Earth station terminals encompass the equipment located both on the ground and on airplanes and ships, meaning that this segment includes air and sea. Communication within space infrastructure can be space-to-ground, ground-to-space, or space-to-space, that is, satellite-to-satellite.

The pirates will likely attack both the satellites directly and the ground/antennae segments. This threat went from theoretical to real in 2023. ORBCOMM, an American satellite telecommunications and logistics company, revealed that a ransomware attack had disrupted its trucking fleet management systems. Customers were unable to use ORBCOMM's products, such as software that enabled the tracking of inventory in transit.[2]

The way the attack took shape is worthy of attention in the context of space piracy. The cybersecurity firm Cybersixgill detected ORBCOMM portal credentials advertised on a popular cybercrime marketplace. Cybersixgill investigators found credentials for what appear to be ORBCOMM's internal portals for sale on the site. Space hackers will likely use similar criminal marketplaces to buy and sell tools required to attack space assets.

Telecommunications Breach

Imagine that you run a major telecommunications business that relies on communications satellites to transmit data across the globe. While you may think of them as space hardware, in reality they are platforms for software that happens to run in space. They are connected to terrestrial computer networks via a radio link.

Like any other network-connected computer running software, your satellites are vulnerable to cyberattack. Your security team has assured you that the software that runs the satellite is protected by firewalls, access control solutions, and other various countermeasures. It better be, you think to yourself. The thing cost millions of dollars to build and launch.

So it is with some surprise that your chief information security officer (CISO) requests an urgent meeting with you. When? Now. Right now. Okay, you say, but this better be important. Before she arrives in your office, you wonder if your billing system has been breached, or some other relatable IT disaster that will cause embarrassment and cost money to fix, but it won't threaten the operations of the business for very long.

That's not the problem, your CISO explains. It seems that hackers have breached the satellite's "command-and-control" (C2) module and are threatening to instruct the satellite to end its orbit and enter Earth's atmosphere, where it will burn up. That is, unless your company coughs up a $50 million ransom, payable in Bitcoin, to make them go away.

Who are they? Well, it's impossible to know for sure, your CISO tells you, but their digital "attack signature" suggests that they are a criminal gang in Russia that operates with the permission of, or even at the direction of, the Russian government and its intelligence services.

Should you pay the ransom? A quick call to your cyber insurance carrier tells you that you're on your own in this mess. Your IT department leadership tells you that the attackers have changed the passwords on the satellite's C2, so they're locked out.

You call your golfing buddy who knows someone at the U.S. Department of Justice (DoJ). "We need to nail these guys," you tell him. "Good luck with that," he replies. As Robert Mueller found when he went after Russian hackers who interfered in the 2016 presidential election, the DoJ can indict criminal hackers who live outside the United States…and have absolutely no impact on their ongoing activities. As long as they don't travel to the United States, they're free to keep hacking our systems at will.

It looks like you're going to have to pay off the space hacking pirates. The problem with this approach, aside from losing $50 million, is that you have no guarantee they'll honor their side of the bargain.

Broadcasting Disruption

If you were the head of broadcast operations for a major cable channel, how would you feel if your channel went off the air for 60 seconds, with no explanation? You'd be in a panic, probably, worrying that your job was on the line as you scramble to figure out what happened so you can explain it to the chief executive officer (CEO), who will be calling you in any second.

The timing of this outage could not be worse. In 24 hours, your channel will soon be airing a massively popular sporting event. The audience will be in the tens of millions. Your channel is poised to rake in hundreds of millions of dollars in ad revenue. If you cannot guarantee a smooth broadcast, you'll be dusting off your résumé in no time.

The good/not good news is that you quickly get an explanation, but it's not anything you're going to feel comfortable sharing with the CEO. Space pirates send you an email describing how they just blacked out your channel. It seems they used a satellite they built and launched via contract. They moved their satellite into position next to yours and jammed its signal. They can do this anytime they want, and there's nothing you can do about it.

After this demonstration of their ability to block the signal, they demand a huge ransom to power down their satellite. If the demand is not met by the next day, all transmission from the satellite will cease. They could even demand a recurring payment, because, for the time being, their satellite is right next to yours, and there is no law enforcement agency or branch of the U.S. government that can do much about it for at least a year, if ever. What can you do about this threat, other than pay the ransom?

An incident of this kind featured into U.S. Air Force Major Andrew Emery's Air University PhD thesis in 2013. In his hypothetical example of space piracy, a nanosatellite moves toward a commercial satellite and "begins broadcasting noise on the uplink frequency of the commercial spacecraft, effectively jamming its target's ability to receive further commands from the ground."[3]

As he further explained, the relatively close range of the nanosatellite means that it can jam its target using only minimal power. The owner of the commercial satellite receives a ransom demand: "unless the company pays the ransom, the offender will continue to hold the company's spacecraft hostage…" Without the ability to upload new commands, the company is at risk of being unable to fulfill its obligations to customer requirements.

Emery bolstered his thesis with a reference to the economic costs of XM Radio's 2007 service outages. The broadcaster went down for 24 hours due to technical problems. Referring to reporting in *USA Today*, Emery stated, "With over eight million subscribers at the time, had each requested a refund of a single day's service it would have cost the company almost $8 million in lost revenue."

Now, you're probably thinking…pirates building a satellite? It doesn't square with your perception of pirates. Aren't pirates a bunch of impoverished desperados zooming around the Horn of Africa in speedboats? Aren't they guys who look like Johnny Depp in *Pirates of the Caribbean*—zany characters with cutlasses and parrots on their shoulders?

Not always. As history and the present day make clear, pirates can operate in highly sophisticated entity structures. They can mask their true intentions as they construct expensive crafts and organize large groups of well-trained operatives. They may even have nation-state backing, or the support of unacknowledged but nonetheless powerful and well-organized nongovernment groups.

Yes, building a satellite and contracting for a launch are big-ticket propositions, but costs are dropping rapidly with nano satellites and advances in rocketry. The payoff is certainly there. If pirates can pull off their threat, they could earn back their investment many times over.

Threats to Commercial Shipping

You keep an antique telescope pointed out the window of your office, which overlooks a major European port. Two centuries ago, this telescope was the only way that the harbor master could tell if a ship was coming into port. A century ago, it was Marconi wireless transmissions from ships that let the port know they were inbound. Today, you have a large flatscreen monitor that shows the positions of dozens of ships as they make their way in and out of the port. It's a satcom display.

You run the port. It's a big job, but you have the assistance of some sophisticated technology. There are more than 50,000 ships at sea or in port at any time. Most vessels have satcom links that are essential to the ship's operations. These satcom links are used to track the status and condition of container ships while in transit. This is especially important when transporting sensitive goods like munitions and hazardous chemicals.

Then, without warning, your satcom display goes black and comes back to life a second later showing just a Jolly Roger skull and crossbones that alternates with an animated video of a gorilla giving you the finger.

This must be a joke, right? You hope it's a joke, something that someone on your team thought would be funny but isn't. You charge out of your office and see, to your horror, that every satcom screen in the place is showing the same animation.

Your phone rings. The caller's voice is distorted by a digital filter. The demand is easy enough to hear, however: you'll be getting instructions by email that describe how your firm, which runs the port, will transmit a hundred million Euros worth of Bitcoin to a crypto wallet owned by an international criminal gang. Then, the pirates will restore the satcom links.

A memo you scanned months earlier pops into your head. Your security team had warned about an event like this, but the matter had seemed so improbable that you let it go. Maybe you could convene a committee to explore the risk and develop an action plan during the coming fiscal year. Otherwise, you would just wait and hope for the best.

Well, the worst, not the best, has come to pass. Your team switches to radio communications with the ships, but it's inefficient and prone to risk. The attack shows how a well-placed hack can put ships' operational integrity at risk. In this case, your port has been impaired, but a pirate band with sophisticated intelligence can identify the vessels transporting the highest-value cargo and simultaneously attack multiple ships with ransom demands.

Other vulnerable but critical satcom-based functions on maritime platforms include but are not limited to:

- Transmitting and receiving ship-to-ship distress alerts
- Transmitting and receiving search-and-rescue coordinating communications
- Transmitting and receiving on-scene communications
- Transmitting and receiving maritime safety information
- Transmitting and receiving general radio communications to and from shore-based radio systems or networks
- Bridge-to-bridge communications
- Notices to mariners
- Maritime/port regulations
- Vessel routing

- Cargo management
- Planned/predictive maintenance
- Radio over IP (RoIP) via walkie-talkie

An attack of this nature is not so hypothetical. The NotPetya hack of the Maersk shipping business in 2017 offered a preview of what may be coming soon. In that attack, hackers implanted the NotPetya ransomware malware in systems that Maersk relied on to run its global shipping operations. The attack badly impaired Maersk's ability to function for weeks. The incident cost more than $300 million to remediate.[4]

A potential variant might be a "Houthis in space" scenario. As the world has witnessed in 2024, the Houthis of Yemen have done much to disrupt global commerce by attacking ships in the Red Sea. A comparable group could, with the right technology, conduct similar acts of disruption against space assets. Driven by politics and funded by a third-party nation, for example, Iran, such a group could cause extensive damage to space commerce.

Aviation at Risk

Everything seems fine as you adjust the yoke and steer the 787 jetliner out over the Atlantic on a routine flight from New York to London. You switch on autopilot and start to relax. The plane will partly fly itself to London. You just have to keep an eye on the controls and manage the landing.

A few minutes later, something very odd happens. The 787's satellite data unit (SDU) goes dark. You try to restart it, but it won't respond to your commands. The autopilot switches off, and you are back to flying the plane the old-fashioned way. Your first officer continues to fiddle with the SDU, but it seems hopeless.

The problem is actually quite a lot worse than you first thought. The Future Air Navigation System, Controller Pilot Data Link Communications, and Aircraft Communications Addressing and Reporting System all cease to function. You are no longer able to use satellite navigation.

Aircraft safety relies on the redundancy and accuracy of onboard systems, many of which are dependent on satellites. Aviation platforms of all types are a very high-value target to pirates. Software security is therefore a mandatory requirement on planes and the ground systems that support them.

Luckily, it's still daytime, and you can still see the coast of the United States out your port side window. Feeling like a first-year cadet in flight school, you use the stick and rudder to turn the big plane around. You head back to New York, radioing ahead that you have a nonfatal emergency but will need to land the plane using visual flight rules.

After calming down the passengers with a watered-down version of the truth, you focus on landing the plane. When you have brought the plane safely back to a waiting gate, you are met by a group of senior executives from your airline, along with a group of FBI agents, who escort you into a conference room and ask you to explain what happened.

At first, you think you're in trouble, but they assure you that you've done nothing wrong. After recounting your bizarre tale, they all share a look. The lead FBI agent then explains that the airline received a cryptic message from a previously unknown group of hackers who said that this incident was a warning—a demonstration of their power to shut off critical aircraft subsystems at any time.

They're space pirates, and they are demanding a massive ransom to allow the airline to fly unmolested. You are somewhat relieved that this is not your problem to solve, but you realize, at the same moment, that your life could be in danger, as well as those of passengers and crews, any time you fly.

Space pirates, working as hackers, could attack satcom infrastructure that enables safe, smooth aviation operations. They can target the ground elements of the infrastructure, which are well defended but still vulnerable to determined, sophisticated attackers. They can affect civilian as well as military flights. By taking over satellite communications, they could bypass authorization mechanisms to access interfaces. Any of the systems could be breached or rendered nonoperational.

A Satellite Breach at a Media Company

Congratulations! You are in charge of operations for a major media company with its own fleet of satellites that broadcast your television programming all over the United States and Canada. So, it is with some excitement that you fly on the corporate jet to Florida to check in on the progress of your latest orbiting profit center.

When you arrive, you can tell that something has gone badly wrong. The chief of the security team takes you aside and explains that you have a big problem: the massive cargo plane that was supposed to deliver the satellite from the manufacturer in California seems to have disappeared.

You're stricken. Did it crash? Your head spins as you try to figure out how to handle a disaster of this magnitude. People might be dead. The company will be out millions. The launch could be delayed by months or even years if you have to build a new satellite. This could affect your job. Turns out, the situation is actually worse than that. The plane, and its extremely valuable cargo, has been hijacked.

Wait, what? While it seems that no one has been harmed, yet, a team of mercenaries stormed the runway when the cargo plane was finishing its loading procedure. Security at the airfield was minimal because who would ever think that someone would want to steal a satellite? The attackers, one of whom was a trained pilot, took the crew hostage, commandeered the plane, and took off.

Where is it? No one knows. You're waiting for a message and ransom demand from these space pirates, who don't even have to go into orbit to threaten a space-based business.

In the meantime, you initiate a prepared emergency response plan. You call your CEO, your general counsel, and your head of risk management and tell them what's happening. They contact the FBI, but there isn't much anyone can do until you hear from the space pirates. Your phone rings. It's a reporter, who's heard what's happening. Should you talk to him?

Hours go by without a word from the pirates. When they call, it's to inform you that your satellite is now in South America. If you want it back, you'll be paying them $25 million. The FBI tells you there isn't a lot they can do without knowing exactly where the satellite is. Your insurance company is suggesting that your policy doesn't cover acts of piracy.

This scenario may sound like something out of a Hollywood blockbuster, but thefts of valuable cargoes are common. Indeed, Latin American criminal cartels routinely seize shipments of oil and sell them on the black market. In Africa, the Nigerian Navy has been credibly accused of stealing oil shipments, as well.[5] It's only a matter of time until an organized criminal group, perhaps with hired mercenaries, steals a valuable space asset.

Attack on a Commercial Spaceport

As this last scenario shows, a satellite doesn't have to be in orbit to be targeted by pirates. Picture this: you are in charge of one of America's many commercial spaceports. It's not a government facility, even though you handle government launches. Your team is going through its exquisitely well-planned process of preparing a satellite for launch.

Then, just as the rocket holding the satellite is mounted under the wing of a specially modified 747 aircraft, a group of mercenaries drives onto the runway and orders all of your personnel to leave the site at gunpoint. Everyone complies, sensibly. And, while you have a few armed guards posted around the spaceport, they are not close enough to the plane to do much, and even if they did, they're overwhelmingly outgunned.

They don't kill anyone, thankfully, but within minutes, they've established a heavily armed cordon around the plane and your rocket. A few minutes after that, they've wired the entire plane and rocket with high explosives. They can now destroy a satellite that took years and millions of dollars to build, in a matter of seconds.

The pirate mercenaries give you a deadline. When they receive word that you have transferred $10 million into their Swiss bank account, they will vacate the spaceport and disappear into the remote mountain wilderness that surrounds it. By the time anyone puts together a suitable law enforcement or military team to go after them, they'll be long gone. In fact, a later analysis of the attack reveals that they flew into a nearby abandoned airfield, most likely from Latin America.

What should you do? You could stall them and try to put together an armed response team that can storm the plane. Of course, that risks everything, and everyone, getting blown to bits. The pirates don't seem to mind. Like pirates throughout history, they appear willing to die in their quest for treasure. This may not be true, but are you willing to test this hypothesis?

A hasty, panicky conference call with the satellite owner confirms that they will pay the ransom. You communicate this to the pirates, who wait until they receive confirmation of the bank transfer before disconnecting their bombs and fleeing the scene.

You've avoided a major disaster, but think about how this episode will reflect on your spaceport's reputation. Will other satellite owners entrust you with their valuable payloads? At a minimum, you're going to need to beef up your security significantly.

Present and Near-Term Risks: Attacking the Space Economy

The $500+ billion space economy will be a very tempting and inviting arena for piracy. Satellites represent just one area of present-day risk from space pirates. Space vehicles, whether manned or unmanned, are a potentially even more serious point of vulnerability for the government entities and corporations that launch them. And the vehicles don't even have to be in space to be targeted by pirates.

Satellites are currently the largest element of the space economy. In the next few decades, however, space operations will likely expand to include the following:

- In-space manufacturing
- In-space satellite/orbital servicing
- Increased unmanned and manned cargo platforms to the International Space Station (ISS) and private space stations
- Space tourism via rockets or balloons
- Unmanned and manned presence on the moon and other planets (e.g. Mars)
- Resource extraction/habitation of the lunar surface
- Cargo from the moon to asteroids to Earth

Each one of these space activities, which will be covered in greater depth in Chapter 3, is a target-rich environment for well-financed, space-tech savvy, and determined space pirates.

A Trip to the International Space Station, Interrupted

What if you were in charge of the launch of a rocket taking people to the International Space Station (ISS)? Or, in the not so far off future, you might be overseeing the transportation of workers from Earth to a lunar outpost. The vehicle that will take them there is, like a satellite, a piece of hardware that depends on software for its safe operation.

Every system on your space vehicle, from navigation to life support, depends on software to function. And all of that software is connected, via network, to computers on Earth. (Even if it isn't supposed to be connected, you'd be amazed what hackers can find exposed to external probing.)

It's also susceptible to "supply chain attacks," wherein malicious actors implant viruses into the software code while it's being developed.

A few hours after a successful launch, your team leader gets a strange email from a hacker with an alias like J0ll1R0Jer that threatens to shut off the life support systems on the space vehicle if he doesn't receive a ransom payment. To demonstrate that he's for real, he actually shuts off the air-circulating fans in the vehicle for one minute. This act causes your space passengers to break out into a sweat, literally, while also making them concerned that something has gone wrong with their craft.

What should you do? Or, perhaps a better question might be "What can you do?" The attack reveals that whatever cyber defenses you had were not effective at keeping the attacker out of the networks that link to the space vehicle. This is a common problem in cybersecurity, but in your case, people's lives are in danger.

If you have a superb team in your security operations center (SOC), you might be able to find the attacker and lock him out of your network. Otherwise, you're pretty much stuck. You can notify law enforcement, or whatever government agency is responsible for your security, but by the time they can do anything, you'll be dealing with a lethal attack.

You're going to pay that ransom. That's the reality of a situation like this.

The only good news in this scenario is that the hacking of space assets is not as easy as hacking computer systems on Earth, for a variety of reasons we will explore in Chapter 5. Still, space assets are exposed to significant cyber risk.

An Attack on In-Space Manufacturing

You run a factory. In space. Sounds incredible, and it is, but at some point in the future, there will be industrial operations in space. It's called *in-space manufacturing*, and it involves the production of tangible goods beyond Earth in space. The goods to be made will be, but are not be limited to, replacement parts for space platforms, novel materials that have never been made on Earth, and experimental hybrid Earth/novel materials and coatings.

Exotic as your job sounds, the day-to-day operations are not that different from manufacturing on the home planet. You've got machines going. You've got deadlines and quotas to meet and logistics to handle.

The difference is that your entire manufacturing process is hundreds of thousands of miles away. It's remotely controlled. You watch your equipment doing its thing on a video monitor, while data about the equipment's functions appear in parallel on a separate set of screens.

Then, after you return from lunch one day, you notice the presence of a robotic device you've never seen before. You check your schedule, but there's no mention of the arrival of new hardware. Nor is there a servicing appointment for this today. What's going on?

It seems that space pirates have arrived. Using a robotic platform they built and launched using a collection of shell corporations on Earth, they're about to steal the extremely valuable goods you've been making.

Their plan is not immediately clear. They might sell it back to you, an idea that makes you crazy. Or, they might sell it to someone else, someone who lacks the ethical compass to avoid buying stolen goods. Whatever it is, the cartel behind the pirates wants a piece of your action.

Bigger picture, this scenario is about how space pirates, using very advanced planning and tech methods, can subtly insert themselves into the in-space manufacturing process. If they are not stopped, they can go beyond theft and produce corrupt products and perhaps destroy the entire in-space manufacturing platform. The space pirates will be able to demand a recurring fee and/or become partners in the company that owns the in-space manufacturing platform.

Obstruction of Satellite Orbital Servicing

As a space entrepreneur, you saw the opportunities in satellite orbital servicing earlier than others. You lined up venture capital, and then private equity partners, to build a sizable business repairing satellites and recharging worn out batteries, among other services. Your company also acts as a platform to provide propulsion systems to rockets and other space transportation platforms. It's a growing sector, one that you are poised to exploit. It includes lucrative projects like cleaning up space debris/dead satellites that have been floating around in space for years. These objects comprise a potential danger to newer space platforms.

Unfortunately, as your fellow space businesspeople have been finding, in-space servicing offers similar opportunities for plunder as in-space manufacturing. Space pirates, via remote control, can gain control of the in-space serving platforms and threaten to destroy not only the in-space servicing platform but also the satellite or other platform being serviced unless they receive a ransom.

You find yourself in a jam. Using a purpose-built, unmanned space vehicle, the pirates approach your servicing vehicle and your client's satellite. The pirates have the ability to destroy your equipment, while also pushing the client's satellite out of orbit. This could result, at a minimum, in a costly process of reestablishing the correct orbit for the satellite. At worst, the pirates could cause the satellite to reenter Earth's atmosphere and burn up.

The attack presents you with a twofold problem. Your own business is in jeopardy, because, make no mistake, an incident like this damages your reputation for safety and security while causing you to lose untold millions in space equipment. It's also a big crisis for your client, who will probably sue you for negligence whatever the outcome. You may want to pay the ransom, rather than asking the client to pony up the money. That, too, could wipe you out.

Hijacking a Space Tourism Trip

You smile at your $250,000 per-ticket guests one last time as they pose for selfies and board your space tourism craft. You smile because as the founder of a space tourism business, things are looking good, in financial terms. It's the trip of a lifetime, part of a growing trend that started with space flights sponsored by Blue Origin and Virgin Galactic that took passengers into space and gave them a once-in-a-lifetime view of space and an experience of low gravity.

Excitement aside, space tourist activities provide several inviting access points for space pirates to hijack these brief flights to the upper atmosphere and extract substantial plunder from the companies promoting these flights. Indeed, with most passengers being ultra-wealthy, space tourism is almost a piracy accident waiting to happen.

One way would be for the space pirates to take control of the operating systems of the launch vehicle and/or its guidance system. Access to these systems by the space pirates could come from personnel inside the space tourist companies who are planted by the space pirates. Or, they could assume control of the life support and related systems of the capsule containing the tourists.

The modus operandi of the space pirates could also occur with another space tourism platform: balloons in space. This space tourism will be even easier for space pirates to hijack because the business model of space tourism via balloon is to create a luxury balloon capsule that will rise 100,000 feet for a six-hour tour of the space horizon. Companies like Space Perspective have received funding to build their first balloon to take a dozen individuals into space.

This platform is also vulnerable—more so than the space tourism platforms used by Blue Origin and Virgin Galactic. Space pirates could insert members as tourists who will physically take over the space tourism platform, analogous to hijackers on planes. Escape could be on a return to Earth with a hostage or via another space capsule built and operated by the space pirates.

Because of the planning involved by the space pirates, including the insertion of "space tourists" to hijack the balloon and guide it to a remote and secure landing site controlled by the space pirates—all experienced personnel encompassing all aspects of the balloon tourist platform—the odds of success and the significant element of surprise will maximize global publicity and plunder via ransom.

In your case, it's a simple system hack, but it's potentially lethal. Fortunately, you get wind of the attack before liftoff. The pirates have threatened that they will go public with their successful hijacking of the vehicle. This threat alone—and the negative publicity it will generate—are enough to make you consider paying a ransom just to make them go away. The last thing you want is for the public to know about your vulnerabilities to pirates.

It's blackmail. They can keep making these threats, and you will have to keep paying if you don't want to be exposed. Your PR person says you should do a planned announcement of the threat to get out ahead of the story. That might work, but it's risky.

Several emerging commercial space station and space tourism ventures will also likely face this risk scenario. For example, Axiom Space is planning to deploy a commercial operated space station, working under a design commission from NASA.[6] Blue Origin and Sierra Space have a plan for a commercial space station, as well.[7] And Hilton Hotels is working with Voyager Space on space lodging.[8]

Longer-Term Risks: Piracy on the Lunar Surface

You've got quite a view. The window is not big, because you're on the moon, but you can see a wide swath of the Sea of Tranquility. Your grandparents could never have imagined such a thing.

You're the on-site manager of a lunar mining operation. Your operation uses highly advanced robots to mine, transport, and upload lunar resources to awaiting cargo ships that will transport coveted resources including water, hydrogen, gold, platinum, magnesium, calcium, aluminum, manganese, titanium, helium, and rare earth metals. These minerals will be used in the next generation of electronic devices. The water and hydrogen will be used for propulsion systems. The operation cost billions of dollars to establish, but it's a viable business that is generating billions in revenues and earnings for your shareholders.

This scenario is admittedly quite a few years out in the future, but it seems inevitable that humans will exploit the riches found on, and under, the lunar surface for business purposes. This exploitation will go hand in hand with establishing an ongoing human presence there. Even today, there are already advanced plans to construct, via remote communications and special materials, structures that will house humans, food/water, and various vehicles, machines, and communications technology/computers on the moon.

Lunar industry and settlement present a scenario comparable to the one facing the buccaneers of the Caribbean in the seventeenth century. To space pirates, human commercial activity on the moon, with its transports from the moon to Earth, will resemble the Spanish galleons taking the riches of the New World back to Spain.

These historical echoes of the golden age of piracy and plunder will be a very powerful attraction to space buccaneers to replicate the success of their forebears. By the time lunar commerce and habitation become established, space buccaneers will have developed many years of experience in successful piracy in space, especially in the technologies needed for successful raids, hacking, and the transportation of plunder back to their own space havens and Earth. They will have also established a network of buyers for their plunders, just as they had in the New World.

As the data and physical riches become pronounced, attacks and extraction of plunder will move from ransom—though this will still be a prominent source of revenue for space pirates—to the actual keeping of the extracted data and physical cargo.

The space cargo ships with physical resources from the lunar surface heading toward their destinations could be attacked via specially designed buccaneer spaceships that will quickly disable the cargo ships via advanced technologies like lasers. Using specially designed robotics and human support, the space pirates extract the cargo from the cargo platform to a space pirate's spaceship. This is a rather futuristic notion, but it is a scenario to consider when designing lunar settlements and transport infrastructure.

Another option for the physical hijacking of space cargo from the lunar surface would be for the space pirates to simply take over the moon and assume control over its resources and its governance.

One scenario of this takeover would be for the space buccaneers to assemble a large, experienced team of space soldiers with experience in space buccaneering. The space buccaneers will deny all data, voice, and video communications going into and out from the moon. They will also be disabling all vehicles and machines on the lunar surface via judicious hacking.

The combination of a complete denial of all networked devices on the lunar surface and overwhelming physical force via the buccaneers, who have completely surprised all the inhabitants, will enable the space pirates to assume control over all lunar activities and quickly take all the inhabitants of the moon as captives.

The space pirates, who have planned this operation for many years, have also ensured they have ongoing access to their own supply chain to replace the provisions used by their captives. The space buccaneers have also made sure they have the skill sets to operate all critical centers on the moon, thus, obviating the need for their captives.

The space buccaneers will have advanced from ad hoc plundering of space assets to establishing a strong and permanent base on the moon and extracting its wealth for themselves. In short, the space buccaneers will be the new governing authority of the moon. Their captives will be bartered for a substantial ransom.

This attack on the moon will, of course, not go unnoticed, and the space authorities will inevitably plan and execute a counterattack. Thus, this space buccaneer takeover of the moon would trigger the first armed conflict in space between the space buccaneers and the space authorities—assuming that such authorities exist, which is not at all clear at the current moment.

The space buccaneers will be heavily armed, experienced in space, and fearless. They will use novel space warfare tactics covering the full spectrum of warfare, from cyber, digital, nuclear, intelligence, and psychological to lasers/explosives. The space authorities, while not having the space warfare experience of the space buccaneers, will have an advantage of numbers and support.

How Likely Are These Scenarios?

Is any of this actually going to happen? It's a fair question, and we don't have a crystal ball, but the truth is that some of these uncomfortable space piracy scenarios could easily happen today. Some of them are already underway, and others could easily happen today with existing technologies.

Others are a little more futuristic, but that future is fast approaching. Consider that *Die Frau im Mond* premiered in 1929, just 40 years before Apollo 11. A lot can happen in 40 years. Who knows where we'll be in space by 2064?

As of the present moment, a group of security researchers demonstrated, using the decommissioned satellite Anik F1R, launched in 2005, that they could breach its cyber defenses.[9] In 2023, *Wired* reported that researchers from Ruhr University Bochum and the CISPA Helmholtz Center for Information Security showed, conclusively, that many satellites in orbit today are vulnerable to attacks on their firmware and a variety of other attack surfaces.[10] They warned that attackers could crash satellites back to Earth.

Symantec reported in 2018 that Chinese hackers, almost certainly working at the behest of the Chinese government, launched cyberattacks on operational satellites using the alias "Thrip."[11] It was not clear if these attacks were successful, but the fact that they allegedly occurred, and were reported on, shows how serious this threat is becoming.

Taking the cyber risk of satellites into account, the U.S. military has responded by commissioning "white-hat" hacking[12] to conduct penetration testing on the U.S. Air Force's Moonlighter Satellite under controlled conditions. The hacking exercise took place at the 2023 DEF CON hacker's conference in Las Vegas. Interestingly, there were two hacker teams at work. One was trying to break into Moonlighter, while a second team was attempting to keep attackers out. The attackers were successful in penetrating the satellite's cyber defenses.[13]

The U.S. military is to be applauded for actively trying to learn where its vulnerabilities are, but it's worth noting that it took five years from the alleged Chinese Thrip attack to this session of penetration testing. This kind of delay is not an auspicious sign for the prospect of defending against space piracy.

Getting Inspired by Science Fiction

Inspired as these scenarios may be, they're nothing compared the rich, extensive works of science fiction that envision piracy in space. Space piracy and crime have long flourished in dozens of books, movies, and TV shows. A small sampling of these works includes *Space Pirates of the Black Nebula*, by Joseph Delmari, which tells the tale of a mercenary spaceship commander pulling off heists across a colonized solar system. *The Terran Privateer*, by Glynn Stewart, is an adventure featuring a space-based pirate/privateer fighting against an alien armada. In David Lee Summers' *Firebrand Legacy*, a space pirate, holding a letter of marque, raids ships belonging to Earth's opponents.

Cosmic Corsairs, a 2020 anthology of space pirate fiction from Baen Books, captures many of the themes we explore in this book. (Corsair is another word for privateer.) In the introduction to the book, titled "Yo-Ho-Ho and a Bottle of Oxygen," editor Hank Davis put forth a nearly identical argument to the one we pose here: "If space travel becomes cheap enough to make a more or less honest buck, it would be cheap enough for less honest (dare I say scurvy?) individuals to hijack it." He also makes the point that, like huge alien monsters that are not scientifically credible, but make for good stories, the idea of space piracy, perhaps not entirely realistic at the moment, is a lot of fun to write.[14]

Many of the stories in the anthology are more or less space versions of classic pirate tales from the high seas—interstellar, interspecies, fanciful, and engaging, but not very current in their outlook. One story, though, "The Barbary Shore" by James L. Cambias, contains a scenario quite close to one that we have contemplated in this book. A space pirate sits on Earth, where he hacks a ship carrying a cargo of helium-3 so it lands where he can steal it.

In films, the 1981 movie *Outland*, starring Sean Connery, stands out as a great example of a somewhat realistic, if futuristic, crime story set in space. In the movie, Connery plays a U.S. Marshal assigned to patrol a mining operation on Jupiter's moon, Io. He encounters a sinister drug smuggling operation and engages in a deadly game of cat and mouse in pursuit of the bad guys. In 1981, the concept of space mining was a very far-fetched idea. Today, it's actively on the drawing board at many space ventures.

On television, the 1960s show *Lost in Space* had an episode called "The Sky Pirate," referring to a wandering space buccaneer who visits the Robinson family and causes trouble. *Star Trek: Voyager* featured an episode titled "The Void," referring to an area of space that was a complete vacuum. No one could leave the Void, and it was empty of anything humans could use to survive. As a result, anyone trapped in the Void had to resort to stealing what they needed to live. Somehow, the crew of *Voyager* convinces all the pirates in the Void to stop stealing and they escape to fight on another day.

Takeaways

Space piracy is a future problem that is starting to show itself in small-scale hacks. However, the probability of space piracy and crime becoming serious issues facing space industry and national security organizations suggests that we should start thinking through possible scenarios today as a way to begin planning effective ways to deal with the issue. Potential space piracy scenarios include disrupting satcom, affecting shipping and aviation, and hacking broadcasting and telecom satellites, with ransom demands made on their owners, kinetic attacks on spaceports, and takeovers of space-based mining and in-orbit business operations. The world of science fiction has been pointing out the potential for such space-based crime for many years, and we can learn from authors' creative minds.

Notes

1. Matt Mills, "How Many Satellites are there Right Now Orbiting the Earth?" ITIGIC, November 22, 2021, https://itigic.com/how-many-satel lites-are-there-right-now-orbiting-the-earth/?expand_ article=1.

2. John Kingston, "Ransomware Attack Hits Orbcomm's BT Series of ELDs; Paper Logs are Back," *FreightWaves*, September 14, 2023, https://www.freight waves.com/news/ransomware-attack-hits-orbcomms-bt- series-of-elds-paper-logs-are-back.

3. Andrew Emery, "The Case for Space Crime: The Rise of Crime and Piracy and the Space Domain," Master's Thesis, Air University, 2013.

4. Nate Lord, "The Cost of a Malware Infection? For Maersk, $300 Million," Digital Guardian, August 7, 2020, https://www.digitalguardian.com/ blog/cost-malware-infection-maersk-300-million#:~:text= For%20Maersk%2C%20%24300%20Million,-by%20Nate%20Lord &text=The%20NotPetya%20fallout%20continues%2C%20 with,following%20a%20June%20cyber%20attack.

5. Stephen Angbulu, "Just In: Military Responsible for Oil Theft in Nigeria – Asari Dokubo," *Punch*, June 16, 2023, https://punchng.com/just- in-military-responsible-for-oil-theft-in-nigeria-asari- dokubo/#:~:text=%E2%80%9CThe%20military%20is%20at%20 the,doors%20at%20the%20Aso%20Rock.

6. Jeff Foust, "NASA Selects Axiom Space to Build Commercial Space Station Module," *Space News*, January 28, 2020, https://spacenews.com/nasa- selects-axiom-space-to-build-commercial-space-station- module.

7. Jeff Foust, "Blue Origin and Sierra Space Announce Plans for Commercial Space Station," *Space News*, October 25, 2021, https://spacenews .com/blue-origin-and-sierra-space-announce-plans-for- commercial-space-station.

8. Kristen Savoy, "Hilton and Voyager Space to Partner on Improving Stays in Space—Designing Crew Lodging, Hospitality Suites for Starlab Space Station," *Hilton*, September 20, 2022, https://stories.hilton.com/ releases/hilton-voyager-space-partnership-starlab- space-station.

9. B. David Zarley, "An Old Satellite was Hacked to Broadcast Signals across North America," Freethink, April 14, 2022, https://www.freethink.com/ space/decommissioned-satellite-hacking.

10. Matt Burgess, "Satellites are Rife with Basic Security Flaws," *Wired*, July 20, 2023, `https://www.wired.com/story/satellites-basic-security-flaws`.

11. Threat H. Team, "Thrip: Espionage Group Hits Satellite, Telecoms, and Defense Companies," Symantec, June 19, 2018, `https://symantec-enterprise-blogs.security.com/blogs/threat-intelligence/thrip-hits-satellite-telecoms-defense-targets`.

12. Joseph Gedeon, "For the First Time, U.S. Government Lets Hackers Break into Satellite in Space," *Politico*, November 8, 2023, `https://www.politico.com/news/2023/08/11/def-con-hackers-space-force-00110919?cid=apn`.

13. Christian Vasquez, "How a Hacking Crew Overtook a Satellite from Inside a Las Vegas Convention Center and Won $50,000," CyberScoop, August 16, 2023, `https://cyberscoop.com/mhackeroni-hackasat-space-def-con`.

14. Hank Davis and Christopher Ruocchio, *Cosmic Corsairs*. Simon & Schuster, 2020.

2

Learning from Piracy's Long and Rich History

The end of the pirate Bartholomew Roberts in an attack by the Royal Naval warship Swallow, which resulted in Roberts' death, 1722 (painting by Charles Edward Dixon)
Source: Archivist/Adobe Stock Photos

Understanding the multi-millennial history of piracy can help us understand the potential for piracy in space. In this chapter, we will guide you through a brief history of piracy. This is a fun subject, though much bloodier than the romanticized pirate stories we get from Hollywood. However, it is also quite instructive for thinking about looming threats in space. Piracy has been around since there has been shipping on the seas. Indeed, it's happening right at this very moment in multiple spots around the globe. Space is merely a new domain for this enduring human criminal phenomenon.

A Brief History of Piracy

The earliest documented instances of piracy were in the fourteenth century, when the Sea Peoples, a group of ocean raiders, attacked the ships of the Aegean and Mediterranean civilizations. The insight of the Sea Peoples, which has been followed by all pirates since and will be adopted by space pirates, is knowing that narrow channels that funnel shipping—or spaceships—into predictable routes have created and will continue to create opportunities for piracy.

For this section of the book, we are deeply indebted to the research and writings of Angus Konstam, one of the world's foremost authorities on the subject. His books *The History of Pirates*, *Pirates 1660–1730*, and *Buccaneers 1620–1700* comprise the basis for this overview.

Pirates captured Julius Caesar. Pirates attacked Rome until Pompey had them removed from the Mediterranean. Pirates—mostly Arab—were a major problem for shipping in the Middle Ages and the Renaissance. This problem lasted until the nineteenth century, especially off the Barbary Coast of Africa.

The Talmud, the authoritative book of Jewish law that was first compiled more than 1,500 years ago, contains a reference to piracy at the time of King David, ca 1000 BCE. The Talmud's *Book of Sanhedrin*, which deals with systems of courts, says, "The Scholars approached King David and said, 'your people Israel need prosperity.' He replied, 'Let them do business with each other.' They replied, 'that would be insufficient.' He replied, 'avail yourselves of raiding camps.'"[1]

Piracy went on to play a significant role in the geopolitical struggles that defined the era of exploration and colonialism that created the modern countries of the Western Hemisphere. For instance, when England captured the island of Jamaica in 1655, pirates had a safe harbor to attack

Spanish shipping and settlements. This piracy was a cause of the decline of the Spanish empire. Comparable forces are at work today in different places around the globe. More on this later in this chapter.

The so-called golden age of piracy in the early eighteenth century was fueled by the long war between England and France, a war that forced some unemployed sailors into piracy. During this era, pirates became active in the North American British colonies, the Mediterranean, the Indian Ocean, and the South China Sea. This was because piracy tends to thrive in places where there is no strong government. The coast of North American countries thus became a haven for pirates. When government power did establish itself in this area, backed by naval forces and a system of courts, piracy went into steep decline.

Privateers

As conventional piracy became less viable, sea captains looked for new ways to make money. One option was to become a privateer. Privateers were men or ships permitted via a "letter of marque" by a government to engage in piracy against enemy ships in wartime. The government, in return, got a portion of the profits. The label was not fixed. A privateer could slide back into basic piracy, and often did.

Privateering enabled countries to attack and disrupt their adversaries' maritime commerce at low cost and without having to divert its naval forces. This strategy proved effective, especially for small countries like the United States in the War of 1812.

Speed and deception were the privateer's main tactics. They lured their prey within range before they could fight back, going for a fast surrender that saved lives and avoided damaging valuable cargoes. Privateers increased their odds of success by utilizing the "wolf pack" group-hunting tactic that would later be used by German U-boats in World War II.

It's tempting to think that privateering would not work in today's connected world, where the actions of a nation-state face tight public scrutiny. However, the perceptive reader might see how modern governments could quietly sponsor space privateers. Later in the book, we will discuss how space privateers can become a major force in the space ecosystem.

Furthermore, the reality is that privateering is occurring right now, in cyberspace. Though cloaked by deniability, criminal cyber gangs are operating out of Russia, disrupting the United States and other Western powers with the blessing, or even direct instructions, from the Russian government. They get to keep the money they steal. The Russian government gets the benefit of a distracted and impaired adversary. It's privateering for the modern age.

Buccaneers

Buccaneers became the predominant form of pirates early in the seventeenth century. They used havens in Caribbean islands, not occupied by the Spanish, as bases to attack Spanish shipping. Originally, they were settlers and would hunt wild pigs and cattle that roamed free on these islands—getting the name "buccaneer," which is an indigenous word for hunter.

Trade soon emerged between the buccaneer settlers and passing ships. The settlements grew, filling with runaway indentured servants, slaves, deserting sailors, and adventurers. These violent men called themselves the "brethren of the coast." Religious refugees, including Jews who were escaping the Spanish Inquisition, added to their ranks, along with actual pirates and criminals. Paid English soldiers, whose skills were well suited to raiding, also joined forces with the buccaneers.

Their settlements soon became bases for Dutch and English "sea dogs" who raided Spanish ports. When France, Holland, and England went to war with Spain, they issued letters of marque to buccaneers, who became privateers as a result.

Buccaneers used tactics of surprise, mobility, and superiority of morale on land as well as sea. A buccaneer might put together a group of attackers and reach agreement on how the spoils would be divided. An advance party would quietly gather intelligence. The buccaneers would anchor their ships at a distance from the target and then attack on land.

Buccaneer plunder was substantial, with a single Spanish warship, for example, netting the buccaneers 120,000 pesos. This was a vast sum at that time. By 1680, a pretty good-sized chunk of Jamaica's annual 750,000 peso income was lost due to plunder. The attacks led to inflation in Spain. By 1700, a combination of inflation and devastation caused by buccaneering raids started the decline of the Spanish empire.

The End of the Buccaneers

The factors that contributed to the end of the buccaneers stemmed from peace treaties that England, Holland, and France signed with Spain. With peace, the buccaneers lost their safe havens and ports as the governors introduced and executed anti-piracy laws. The peace treaties also dried up the letters of marque, so privateering wound down at that time. With no more cover of letters of marque and safe havens critical to all successful buccaneers, most buccaneers turned to legal activities or became pirates.

American Privateers

Privateering was common during the American Revolution, with ports in New England serving as privateering bases. Salem, Massachusetts, for instance, hosted more than 150 privateers. Privateering led to industry in such ports, with shipbuilders competing to create innovative vessels. The ports also provided financial infrastructure for privateering activities. The money was certainly good. For example, the ship America, a privateer, earned more than $600,000, equivalent to $12 million today.

Privateering was essential for the revolution because hard currency was rare, and the young United States had no reserves. With a volatile economy, inflation was a problem. Privateering ports were among the only places where hard currency was entering the country. It enabled the purchase of commodities needed for the survival of the new country.

The United States was home to two of history's best-known privateers: John Paul Jones, who went from being a ship's master in the British merchant marine to joining the Continental Navy in 1775, and the French-born Jean Lafitte, who led pirates who attacked Spanish ships and sold the plunder to Louisiana merchants.

Piracy, Geopolitics, and Finance in the Rise of the British Empire

It's worth digging a little deeper into the synergy between piracy, geopolitics, and finance that was a factor in the rise of the British empire. Again, while this is an interesting bit of history, we are bringing it up because it will inform our thinking about piracy in the modern world and space.

Mark Hanna, a history professor at the University of California, offers an excellent analysis of this phenomenon in his book *Pirate Nests, and the Rise of the British Empire, 1570–1740*. As Hanna explains, the defeat of the Spanish Armada in 1588 was partly due to the efforts of private men-of-war. This definitive battle in the decline of Spain led to English sea captains becoming less scrupulous about attacking Spanish ships, as well as Dutch ships carrying Spanish goods.[2]

He adds, "High seas piracy flourished only with the active support of accessories on land, yet land crimes came under the purview of common law. Local authorities, however, fought this encroachment of crown prerogative from the very beginning."[3]

Indeed, Hanna goes on to share how England's legal system was ideally adapted (or maladapted, if you will) to encouraging piracy. In the seventeenth century, men found guilty of crimes against subjects of England's allies were seldom brought to justice overseas. If they had been found guilty in London, they would have faced trials. He says, "Piracy was enhanced due to a legal pluralism that existed in England which saw the efficacy of several legal systems that complimented and competed with one another."

In Hanna's view, tension between civil law, which was prevalent in London, and English Common Law, which was based on local tradition and precedent, made it extremely difficult to prosecute pirates. He says, "Common law was often inadequate in dealing with crimes on the high seas. Common law-courts often failed to prosecute pirates because jurors rarely witnessed pirate attacks themselves, and were often socially and financially tied to the defendants."

This jurisdictional conflict resonates with some present-day legal obstacles to combating piracy. For instance, Somalia and Mexico have laws against theft and piracy, but prosecution of such crimes in those countries is uneven, to put it mildly. Russia, too, has a robust legal system that appears to look away as the Wagner Group engages in illegal acts abroad. The story of England's jurisdictional issues also foreshadows potential barriers to bringing space pirates to justice. Law enforcement on the lunar surface, for example, might be quite different from its corollaries on earth.

This legal ambivalence toward pirates carried over to England's American colonies. For example, William Penn, founder of the Pennsylvania colony, was aware of pirates in his city who were, as Hanna puts it, "notorious for committing astounding acts of piracy against Muslim pilgrimage vessels in the Indian Ocean." Some had even married local Quaker women. However, Penn's personal lawyer in England, David Lloyd, who served as the colony's attorney general, refused to try known pirates for their crimes.[4]

Robert Quarry, the crown appointed vice admiral of Pennsylvania, reported that locals in colonies surrounding Delaware Bay had, "entertained the pyratts, convey's them from place to place, furnished them with provisions and liquors and given them intelligence and sheltered them from justice."[5]

Hanna then explores the complex, mutually beneficial—if hidden—relationships between pirates, investors, and commercial/financial markets. These types of relationships exist in modern piracy and will likely play a role in space piracy.

For example, the landed gentry in the area surrounding British ports at the end of the sixteenth century benefited financially by supporting pirates. Their support included provisioning pirates with ammunition and other supplies they needed to pursue their illegal missions.[6] In return, the pirates provided access to markets that were closed to English traders who were not part of Crown monopolies like the East India Company or the Royal African Company.

Offering a case example, Hanna describes how Sir John Killigrew, vice admiral of the Killigrew family, governors in Cornwall, utilized a network of family connections in Wales, Dorset, and Munster in Ireland to fund dozens of pirate crews who stole from ships flying the flags of France, Scotland, and Holland.[7]

The financial returns from these "investments" were substantial. In the years following the defeat of the Spanish Armada, English pirates netted a haul of about £400,000, which was equivalent to more than 10% of all English imports at the time. For comparison, the sum total of crown revenues in 1588 was £392,000.

Plunder went hand in hand with English efforts at colonization. Take Robert Rich, the Earl of Warwick, who inherited a fleet of ships he used for private warfare as a driving force in his colonial investments. Rich, working with a tight circle of family, friends, allies, and servants, invested in or was politically connected to numerous colonial ventures. These included the

Virginia company, the Somers Isles Company (Bermuda), the Guinea Company (Africa), and the Plymouth Company (New England).[8] His private navy shipped people and goods around, while also defending his ventures from foreign militaries and other pirate bands. The Rich story offers an example of the sometimes blurry line between pirates and private, mercenary armies—a line that continues to be blurred today.

Piracy also had a role to play in the balance of trade that affected the financial markets and broader economies of England and other colonial powers at the time. In that era, there was little of the currency "float" that we take for granted in the modern era. Imports had to balance exports to maintain equilibrium. In other words, gold and silver were a nation's wealth, despite the development of paper style currencies. Precious metals were the primary medium of trade.

This economic system was driven by a concept known "bullionism," which itself created a continuous, arguably obsessive drive to increase a country's holdings of gold and silver. Piracy enabled success in this compulsion to amass metal wealth. In 1575, for instance, as the British crown faced what it perceived as a currency crisis, Queen Elizabeth I used silver stolen from Spanish treasure ships by private men-of-war (i.e. pirates) to maintain the value of the pound sterling.

This history shows that piracy, and the resulting traffic in stolen goods and flows of illicit money, can be an open secret in commercial and financial markets. Surely, everyone who needed to know was well aware that the silver that flooded into English coffers was stolen from Spain by pirates. But, did they care? For the most part, they didn't, even if some people made official noises of complaint about these practices. Compared to poverty and chaos, however, ignoring a little piracy was a perfectly fine, arguably even patriotic thing to do. This issue continues today and will likely play a role in piracy in space.

Piracy Today

The world is witness today to much piracy. It lacks the swagger and romance of a blockbuster movie, but it is nonetheless a serious problem. Some experts put the cost of piracy at $25 billion per year.[9] Examples abound, but for our purposes, a look at piracy at the Horn of Africa and the Gulf of Mexico will be instructive for thinking about potential piracy in space.

Piracy at the Horn of Africa

In 2008, Somali pirates hijacked the Sirius Star, a fully loaded oil tanker, 450 nautical miles off the coast of Kenya. The incident was a shock to the crew and the ship's owners, who thought the vessel was well out of the piracy danger zone in the Gulf of Aden. They were mistaken and found themselves confronting a $25 million ransom demand to release the ship and its crew.

This actual payment eventually got settled at $3 million after months of negotiation, during which the Sirius Star's crew was held hostage in the ship's cramped quarters. The saga of the Sirius Star demonstrated, to many in positions of authority, that something needed to be done about rampant piracy at the Horn of Africa. Looking at the map, one can see how ideal Somalia is for pirate operations. It juts out into the Arabian Sea, south of Saudi Arabia, and the Red Sea, pathway to the Suez Canal. More than 25,000 ships, many of them laden with precious cargoes from Asia— including some 18% of the oil consumed by the United States and Europe[10]—move along 1,800 miles of Somali coastline in an average year.[11]

By the early 2000s, this lucrative daily procession of maritime prizes was steaming past a failed state. The government of Somalia had little control over its territory. The country was awash in seemingly endless violent conflict. Its economy was in a state of collapse, with annual per capita income in Somalia at that time was less than $300 per year. Such conditions left tens of thousands of young men with no prospects and some strong incentives to join pirate gangs, or the land-based gangs that ran them.

The toxic synergy of a failed state and a steady stream of nearby, poorly guarded multimillion-dollar cargoes led almost inevitably to a burst of piracy. Most Somali pirates operated out of the Port of Bosaso, using small boats and AK-47 rifles to board and take control of merchant ships in the Gulf of Aden. Between 2007 and 2012, pirates captured more than 200 vessels in the seas off the coast of Somalia. Sticking, for the most part, to the "pirate's code," which avoids violence and murder, ransom payments were excessive, topping $200 million in 2010 alone.[12]

In response, a set of coalitions took shape comprising governments, militaries, and private companies to combat the scourge of piracy at the horn of Africa. The fine-grained details are beyond the scope of this book, but some highlights include the U.S. Central Command's Task Forces 150 and 151, which patrolled the high-risk area (HRA) off the Somali coast, as well as Operation Atalanta, which involved cooperation between European governments and the shipping industry.

These efforts have been effective. There has not been a major ship hijacking in the area since 2012. The success shows that it is possible to mitigate the risk of piracy. It's not a cheap proposition, though, and any project that brings together entities like the U.S. Navy and Chinese Navy for years on end is going to be complicated to carry out. But, it works, offering lessons in how nation-states can come together to combat piracy in space.

Piracy never stops, however. Though piracy is a global phenomenon, it tends to be concentrated in areas where many high-value cargoes pass through a geographically narrow space. In the Gulf of Guinea, for example, pirates are targeting oil tankers from Nigeria and Equatorial Guinea. According to the International Maritime Bureau (IMB), there were 19 incidents of piracy in this area in 2022. This is down from 35 in 2021 and a high of 84 incidents in 2020.[13]

The IMB tracks piracy in the Singapore Straits and Callao Anchorage, a busy area near the port of Callao in Peru. In the Singapore Straits, piracy is on a growth trajectory, with 38 incidents in 2022, up from just three in 2018.

Overall, though, even if the number of incidents is dropping, per IMB data, the results are still disturbing: 2022 saw 115 incidents of piracy on the high seas, in which 95% of vessels were boarded. Forty-one crews were taken hostage. Two were kidnapped. Six were assaulted. One was fired upon. The IMB credits cooperation between coastal authorities for the decline of piracy, but shipping in these zones continues to come with risk.

Thefts of Oil Cargoes

Stealing oil cargoes is an activity that can help us better understand piracy today and its likely practice in space. Oil is today's equivalent of Spanish galleons loaded with gold and silver. A large oil tanker can carry up to 345,000 barrels (bbl) of oil.[14] At $100 per bbl, which was the average Brent Crude price in 2022, that tanker is carrying about $35 million worth of oil.

Such rich targets become attractive not just to gang members in Zodiac boats. Corrupt governments may also get in on the action. In Nigeria, for example, it's an open secret that the Nigerian military diverts oil tankers and sells their cargoes for the personal gain of senior officers. Author Steve Coll discusses this type of government-sanctioned piracy in *Private Empire*, his profile of Exxon-Mobil's global geopolitical entanglements.[15]

More recently, the Nigerian politician Mujahid Asari Dokubo, lashed out against that country's government, saying, "The military is at the centre of oil theft and we have to make this very clear to the Nigerian public that 99 per cent of oil theft can be traced to the Nigerian military, the Army and the Navy especially."[16] Interestingly, however, Dokubo has pledged to support the Nigerian government in its efforts to purge the military of those culpable for oil thefts.

Organized theft of oil cargoes also goes on in the western hemisphere. In the Gulf of Mexico, for example, the Mexican Navy logged more than 300 incidents of theft of petroleum shipments and oil drilling equipment between 2017 and 2019.[17] The details of who's who and who's doing what are not entirely clear, which is common in situations with widespread criminal activity, but experienced observers are fairly certain that a large criminal syndicate is behind the thefts.

The thieves pose as fishermen or even Mexican law enforcement personnel. They capture oil or diesel shipments, apparently working on the payroll of the criminal syndicate, which operates like a cartel. Indeed, the syndicate may be part of one of the major criminal cartels operating in Mexico, with oil piracy serving as a source of revenue alongside drug trafficking and human trafficking.

According to a reporting by Fox News, 40% of Mexico's oil is being distributed illegally. The article said, "One [Mexican] intelligence official surmised that cartels with connections to the pirates and oil thieves even have operated entire gas stations with their illicitly extracted goods."

A portion is believed to leak back into the U.S. black market, too. The article went on to say, "But, the U.S. probably isn't the biggest end consumer. The majority of the stolen oil would be utilized locally, also moving to Venezuela and Colombia in the Latin America region…even potentially going as far afield as China and the surrounding region."[18]

Lessons to Be Learned from Old and New Modes of Piracy

Reflecting on old and new modes of sea piracy, piracy's relationships with the worlds of geopolitics and finance and the relative success of anti-piracy efforts can enhance our thinking through the likelihood and manner of space piracy. We will return to these themes more than once later in the book. For now, though, we can look at piracy's past and present and learn lessons that will help us prepare for space piracy.

Piracy Is Endemic to the Human Condition

For one thing, the similarities between piracy today and piracy in the seventeenth and eighteenth centuries suggest that piracy is simply endemic to the human condition. Where valuable assets must transit tight, hard-to-secure passages, people will try to steal them. And those people will invariably organize themselves for maximum effectiveness and profitability.

There's no reason to imagine this won't happen when valuable cargoes transit hard-to-secure passages in space and space connection points on Earth. It's not a one-to-one analogy, because space is different from the high seas, but the patterns of criminal conduct and organization are likely to be quite similar.

Piracy and Geopolitics

Another similarity between piracy in the age of sail and piracy in the era of container ships is the fluid interplay between piracy, pirate groups, and geopolitical entities. Old-time pirates often functioned as unofficial extensions of nation-states. Acts of piracy were part of broader military strategies and wars. Their plunder helped fill the coffers of countries like England, while depriving their enemies of naval ships and treasure. In Somalia, sea piracy funded local warlords in their efforts to best one another on land.

History and the present illustrate well that sea pirates cannot function without support on land. In the old days, that meant England providing sanctuary, either officially or off the books, to pirates in places like Jamaica.

Today, piracy in Somalia was possible, or even actively encouraged, by forces in a failed state. In Nigeria, if reports are to be believed, it appears that people in positions of authority are profiting personally from oil piracy. One has to assume that the government of Nigeria is aware of this and is perhaps involved in the plunder.

In space, the interdependency between nation-states and space industry will likely produce active, if stealthy, relationships between national governments and space pirates. Alternatively, space pirates may work at the direction (or under the protection) of powerful nonstate actors, such as the Russian Wagner Group of mercenaries.

We will explore the idea of mercenary groups playing a role in space piracy throughout the book, but for now, it's worth looking at some recent events and what they can tell us about likely moves in space. The Wagner Group, a mercenary army with close ties to the Russian government, is active in Africa, for example, serving as a plausibly deniable but obviously connected arm of Russian foreign policy on that continent. They are a powerful force in Africa, helping achieve Russia's economic and policy aims in the region. Observers there have noticed that the Wagner Group is displacing France as the dominant foreign influence in Central Africa. This move is creating an epochal power shift after decades of "France Afrique."[19]

So it may be in space. Russia has one of the world's most advanced space capabilities. It would seem inevitable that the Wagner Group, or some equivalent mercenary organization, will play a role in Russia's efforts to colonize space and develop space industry—just as it has in Africa. Such a group might also leverage space piracy to the benefit of Russia in its terrestrial wars, too. And, as is the case in Africa, who is going to stop the Wagner Group from doing what it does? They could even conceivably turn on their sponsor, a possibility suggested by the events of the Ukraine war. The world's militaries don't seem to be interested. In space, it will be even more challenging to confront a mercenary threat.

Piracy and Commercial and Financial Markets

Piracy, past and present, creates mutually beneficial synergies with commercial and financial markets. In seventeenth-century England, members of the merchant class, such as shipbuilders, profited from piracy. Many commercial enterprises in that era knowingly traded in pirated goods. The Crown and Royal Exchequer knowingly trafficked in pirated gold and silver. Protestations of illegality aside, the finances of pirates were well enmeshed with the finances of the state and commercial interests.

Similar dynamics exist today. When pirates hijack an oil tanker, the business entity that buys the oil must have a pretty good idea that it didn't come from a legitimate source. The transaction occurs, nonetheless. Thus, we see piracy integrating itself into the "real" modern economy, so to speak—including the many commercial enterprises that support the shipping, transport, and processing of stolen oil. Banks are not far behind, likely extending credit to facilitate trade in stolen oil.

It would be naive to imagine that such pirate-to-legitimate market interplay will not take place in space. Because space exploration originated as a government-sponsored activity, replete with patriotic propaganda and the laurels of national honor, it is hard for some people to picture space as a place where criminals prosper. Fiction, such as sci-fi movies, have little trouble with this idea, but still, the notion of the all-American heroic astronaut transforming into a brigand feels false to Americans. We better get used to it, though.

We believe that corporations and their partners in the financial world will find ways to do business with space pirates. In some cases, they may be genuinely ignorant of the true source of the transactions they're profiting from, but more probably, they will be aware that they're engaging in a black market that's an open secret and either pretend it isn't happening or actively cover it up.

The Successes and Challenges of Anti-piracy Measures

The past and present provide instructive examples of how anti-piracy measures can work. In the seventeenth and eighteenth centuries, geopolitical changes took away many pirate safe havens. In addition, concerted military efforts, such as the U.S. Navy's confrontation with pirates off the coast of North Africa, showed that it is possible to beat back piracy by force. More recently, the success of multilateral forces working in concert to patrol the Horn of Africa demonstrates that organized programs from well-funded actors can reduce or eliminate piracy.

The costs and international focus required to combat piracy are not insignificant. It's also impossible to do it everywhere. Piracy springs up in places like the Gulf of Guinea once it's been stopped at the Horn of Africa.

Some of these methods might work in space, but the challenge is going to be a lot deeper. For one thing, the "space" in space piracy is quite a lot bigger than any maritime zone we've ever dealt with. The danger zone at the Horn of Africa is 2.5 million square miles. The spherical space surrounding Earth at low earth orbit (LEO) is thousands of times larger. The spherical space between Earth and the moon is orders of magnitude bigger than that. Patrolling all this space will not be so easy, and the vehicles required for such patrols are more complex, rare, and costly than any navy ship.

Additionally, combatting space piracy means fighting across multiple spatial dimensions. While piracy on Earth occurs at sea, with support on land, space piracy will take place in outer space, terrestrial space, and digital "inner" space. A space pirate might hijack a transport carrying valuable metals in transit from the earth to the moon, but he can land pretty much anywhere on the earth's 196 million square mile surface. (Try patrolling that.) And then, he can move the funds he receives for his stolen cargo in the world's vast, opaque digital financial system—or worse, the nearly untraceable cryptocurrency economy. As with the other ideas just discussed, we will return to this in greater detail later in this book.

The Bigger Picture: Do We Really Understand How Much of the World Works?

Preparing for space piracy forces us (or should force us) to adopt a more realistic view of how the world functions. Not everyone will agree with this perspective, but we want to suggest that, for most of us, our understanding of law, crime, and nation-states is not accurate. We see the world as we wish to see it...the way our major media and political entities want it to be.

This idealized vision of the world holds that each country is a functioning, effective Westphalian state that is in control of its population and borders. Bound by international laws and agreements, these Westphalian states participate in an international community that honors treaties and diplomatic solutions to most problems. Occasional military conflicts are conducted in accordance with the conventions of war. Each country has a criminal justice system, that while imperfect, is committed to bringing criminals to account for their bad deeds, often in cooperation with law enforcement agencies in other countries.

A passing glance at the state of modern piracy and equivalent acts of brazen, global crime and terrorism should put this fantasy to rest. Yes, much of the world tries to function like this, but even the "good" countries are riddled with international criminal activity that eludes easy mitigation, or even detection. This is the world where pirates thrive. It's the world where space pirates will find a welcome home.

To understand this alternate world, a world we very much live in, we are indebted to the work of Professor Margaret Sankey of the Air Force War College. She is the author of the 2022 book *Blood Money: How Criminals, Militias, Rebels, and Warlords Finance Violence*. The book does an excellent job of mapping the connections between criminal enterprises and violent nonstate actors (VNSAs), such as terror organizations. Sankey lays out the byzantine, interlocking worlds of terrorism, criminal organizations, militias, and their (un)willing partners in the "legitimate" spheres of banking and global corporate operations.

The book prompts a thought exercise. Imagine you're watching the news. The first segment concerns "separatists" in some region of the world perpetrating a terror attack. The second item is about a British banker convicted of money laundering. The third item tells of a massive ransomware attack that netted millions for some anonymous hacker. You could be forgiven for not realizing that these are all, in fact, part of a single scheme. The news doesn't know it. Policymakers barely know it, or if they do, they haven't figured out a way to deal with it.

This is the world Sankey explores. Her book contains a number of fascinating revelations, starting with the fallacy that terror attacks and other forms of "asymmetrical warfare" are cheap for the attacker. While yes, an explosive vest might cost a thousand dollars, the organization that planned the attack and recruited the suicide bomber may actually need tens of millions of dollars a year to operate. Sankey lays out the political and bureaucratic realities of running an international terror group. They have personnel departments, accountants, media arms, and so forth. These are big-budget operations, and the money has to come from somewhere.

The money comes mostly from criminal activities, according to Sankey. As she explains, affinities between criminal groups and VNSAs—for example, between Hezbollah and Lebanese expat criminals around the world—create the fundraising synergies required to keep the VNSA in operation.[20]

This synergy comes with a host of problems for everyone involved. The VNSA may want notoriety. Indeed, its operations may be driven by a desire for global media attention. That's exactly what a criminal group doesn't want. They would prefer to operate in the shadows. Pirates may or may not want to remain out of sight. They tend to be visible, but they usually want to mask their identities.

She examines the major areas of criminal–VNSA collaboration, including fraud rings, smuggling and counterfeiting, and piracy. Anywhere there can be arbitrage of commodities or goods, there is the opportunity for this kind of pirate activity. For example, manufacturing fake versions of popular consumer products and smuggling them into the United States or European Union provide opportunities for profit that can fund VNSA projects. There's a reason such fakes are called "pirated goods." Domestic cigarette smuggling to evade taxes is another example she provides.

Sankey also delves into the murky crossover zone that connects the licit and illicit economies. She discusses the problems of "deviant globalization," wherein the commercial and communication networks that exist to exploit cheap labor and resources in the third world are harnessed to deliver money and stolen goods to the West. Deviant globalization will likely play a role in space piracy.

Her analysis of the weaknesses in the prosecution of these entities aligns with some of our concerns about policing space piracy. For example, one of the biggest issues, as she explains, is the "silo" approach taken by law enforcement and policy bodies. Terrorism is one category of enforcement. Crime is another. They should ideally be pursued jointly, or at least in a more integrated process. Certainly, a silo approach will not work with space pirates.

Takeaways

Piracy has been around for millennia. It seems to be part of human existence. When valuable cargoes transit narrow passages, pirates will try to
steal them. Piracy is often connected to wars between nation-states, which
use pirates to disrupt and impair their enemies. This was the case in wars
between England and Spain in the seventeenth century, leading to a "golden
age" of piracy in the Caribbean. Piracy exists today, with ship seizes at the
Horn of Africa in the 2000s a notable recent example. However, thefts of oil
cargoes of the coast of Africa and in the Gulf of Mexico show that piracy
never really stops. Furthermore, piracy has long been enmeshed in nations'
financial systems, such as in seventeenth-century England. Such connections persist today, even if they are hard to see. It is possible to deter or end
piracy, typically through international coalitions like the U.S.-led Task
Force 151 off Somalia. These efforts are costly and cumbersome to manage.

Notes

1. Talmud, *Sanhedrin*, 16a.
2. Mark G. Hanna, *Pirate Nests, and the Rise of the British Empire, 1570–1740*. The
   University of North Carolina Press, 2017.
3. Hanna, *Pirate Nests*, 29–30.
4. Hanna, *Pirate Nests*, 1.
5. Hanna, *Pirate Nests*, "Introduction."
6. Hanna, *Pirate Nests*, 9.
7. Hanna, *Pirate Nests*, 29.
8. Hanna, *Pirate Nests*, 67–68.
9. Freda Kreier, "Piracy at Sea is Waning—But Hotspots Remain," *Nature*,
   January 17, 2024, https://www.nature.com/articles/d41586-024-
   00124-6.
10. Reuters, "Factbox: Somali Pirates Risk Choking Key World Trade Route,"
    Reuters, November 14, 2008, https://www.reuters.com/article/
    us-somalia-piracy-aden/factbox-somali-pirates-risk-choking-
    key-world-trade-route-idUSTRE4AD5HP20081114.
11. Peter Cook and RADM T. McKnight, "The Decline of Maritime Piracy in the
    Horn of Africa," Jewish Policy Center, 2019, https://www.jewishpoli
    cycenter.org/2019/10/10/the-decline-of-maritime-piracy-in-
    the-horn-of-africa/#:~:text=Even%20though%20they%20
    operate%20independently,has%20decreased%20in%20the%20
    region.

12. McKnight, "Horn of Africa."

13. "Sustained Efforts Needed as Global Piracy Incidents Hit Lowest Levels in Decades," ICC Commercial Crime Services, January 12, 2023, https://www.icc-ccs.org/index.php/1324-sustained-efforts-needed-as-global-piracy-incidents-hit-lowest-levels-in-decades#:~:text=The%20ICC%20IMB's%20annual%20report,where%20incidents%20continue%20to%20rise.

14. Mason Hamilton, "Oil Tanker Sizes Range from General Purpose to Ultra-Large Crude Carriers on AFRA Scale," U.S. Energy Information Administration, September 16, 2014, https://www.eia.gov/todayinenergy/detail.php?id=17991#:~:text=A%20GP%20tanker%20can%20carry,8%2D14.5%20million%20gallons.

15. Steve Coll, *Private Empire: ExxonMobil and American Power*. Penguin Books, 2013.

16. Angbulu, "Oil Theft in Nigeria."

17. "Mexican Pirates Posing Greater Risk in Gulf of Mexico to Oil Workers, Tourists," *Hellenic Shipping News*, October 3, 2019, https://www.hellenicshippingnews.com/mexican-pirates-posing-greater-risk-in-gulf-of-mexico-to-oil-workers-tourists.

18. Hollie McKay, "Mexican Pirates Posing Greater Risk in Gulf of Mexico to Oil Workers, Tourists," *Fox News*, September 30, 2019, https://www.foxnews.com/world/mexican-pirates-gulf-mexico-oil-workers-tourists.

19. Justin Ling, "Russian Mercenaries are Pushing France out of Central Africa," *Foreign Policy*, March 18, 2023, https://foreignpolicy.com/2023/03/18/russian-mercenaries-are-pushing-france-out-of-central-africa.

20. Margaret Sankey, *Blood Money: How Criminals, Militias, Rebels, and Warlords Finance Violence*. United States: Naval Institute Press, 2022, 440.

3

The Coming Multitrillion Dollar Space Economy

AI-generated rendering of a commercial space vehicle
Source: lililia/Adobe Stock Photos

This book, so far, is based on a major assumption, which is that pirates will find treasures worth stealing in space or from terrestrial support systems for space industries. Is this assumption correct? As of today, we can offer a partial "yes." There is currently an active, sizable, and growing space industry. It mostly involves satellites, but it appears that much, much more is on the horizon.

It is impossible to predict the future size or nature of space-based industries with any accuracy. This limitation applies to both the size of businesses that make money in space and the time frame of their realization. What we think will happen in 20 years might occur a century from now.

What does seem true, however, is that the space industry has ambitious plans on many drawing boards. Visions for businesses in space range from expansion of the existing satellite industry to space-based electrical generation and lunar industries, including mining, space tourism, and asteroid mining.

These sectors fall into four broad categories. Some are entirely terrestrial in nature. For example, the aerospace industry that supports space commerce is part of the space economy, as are launch facilities, space ports, and so forth. Earth-to-space commerce is the second category. This market, which is in its infancy, involves businesses that generate revenue by selling products made on Earth to entities that operate in space. The third is space-to-Earth, in which products or commodities from space are sold on Earth. The fourth is space-to-space, which doesn't exist yet and is the most futuristic of the categories. Space-to-space industry is envisioned as commerce between corporate entities that operate entirely in space.

How big will these industries be, and what is the timeline for their projected growth? These are not easy questions to answer, but one thing that's certain is that space businesses will exist only if they can justify the required investments. For context, know that private financing for space ventures topped $15 billion in 2021.[1] Considering the costs of space exploration, which are high despite their downward trajectory, profitable space industries will have to be measured in trillions of dollars.

Near-Term Space Industry Market Projections

The space industry today is a known quantity. Most experts peg it at around $450 billion in annual revenue. That's up from $280 billion in 2010, which is an impressive rate of growth. According to research from McKinsey & Company, the industry is on track to reach $1 trillion by 2030.[2]

McKinsey, which presented this research at the World Economic Forum in Davos in 2023, cited a number of factors to explain its space growth forecast. They look at private funding of space ventures, which is growing rapidly, as is the number of space startups in the space economic ecosystem. They look at growth, since 2010, of active satellites, satellites launched, and launches per year. Active satellites, for instance, shot up from 1,000 in 2011 to 4,850 in 2022. With this quintupling of satellites in orbit in 10 years, it doesn't seem a big stretch to predict, as McKinsey does, that there will be more than 15,000 aloft by 2030 (see Figure 3.1).

In parallel, McKinsey forecasts steep declines in the costs of commercializing space. The cost of a satellite, for example, which could be as high as a billion dollars a decade ago, has dropped precipitously, to around $100,000 for a small low earth orbit (LEO) satellite today—a figure that's expected to be cut in half by 2030.

The space sector has come a long way and seems poised for future growth.

Projections for space activities

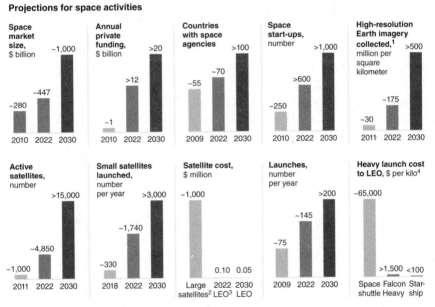

FIGURE 3.1 Space industries are poised for growth across multiple sectors, with the overall market size projected to surpass $1 trillion by 2030.

Source: McKinsey & Company / https://www.mckinsey.com/featured-insights/sustainable-inclusive-growth/chart-of-the-day/a-giant-leap-for-the-space-industry

The cost to launch a kilogram of payload into LEO has experienced a similar, exponential reduction over the last decade, with further predictions of lower cost coming in the next few years. This drop in costs, which is expected to catalyze much in the way of space industry, is largely the result of reusable rockets, such as those created by SpaceX.[3]

The research firm of BryceTech compiled a breakdown of the space economy at the current moment. Their 2022 study puts the value of space commerce at $384 billion, of which $113 billion represents satellite services. Of that, a full $82 billion is revenue for television satellites. BryceTech estimates that the space industry spends $145 billion per year on space equipment, both on the ground and in orbit. The remaining $101 billion spent in space comes from government budgets.[4]

The combination of increased capital flows into the space economy and lower costs will likely synergize to create a larger, more vibrant private space industry than has ever existed before. The business will join existing sectors, such as telecommunications, with extensions of satellite-based businesses and satellite support services. Because these projections are based on growth of known markets and improvements in known technologies, one can look at the prediction of a trillion-dollar space economy by 2030 as being fairly reliable.

Longer-Term Space Industry Space Market Projections

Trying to get a sense of the size and makeup of the space economy in the years after 2030 is a bit more challenging. It's not a wild guessing game. The most probable developments in space from, say, 2030 to 2040, will be based on growth in existing, proven markets, such as telecommunications that utilize existing technologies. Or space will be an environment for advances in existing industries like pharmaceuticals, which can do things in zero gravity that are impossible on Earth. In business terms, the space economy of the next 20 or so years will be evolutionary. It will involve incremental growth and knowable progress in space technology. The specifics, while impossible to predict accurately, are based on current realities.

That said, hype and overactive imaginations have the potential to distort projections of space industry. This is true of any technology-driven commercial category, but space has a particularly bad case of it. From *Die Frau Im Mond* to *Star Trek* to *Star Wars*, fantasies of space explorations have inflamed the minds of men and women. And, to be fair, space is exciting, arguably the most exciting field of all. But, reality is still reality. It's essential for the overeager prognosticators to take a deep breath and think calmly about what is likely in space and what is not happening in the coming half century or so.

New Propulsion Technologies

The way that rockets have gotten out of Earth's gravitational field has not changed much in the last 60 years. There have been incremental improvements, but the basics remain the same. Rockets have "escape propulsion" systems that burn chemicals in liquid or solid forms. A growing industry is taking shape, however, for new propulsion systems to power "in-space" and "deep-space" missions.

In-space propulsion is about getting spacecraft from LEO to higher orbital positions, such as the position of the International Space Station (ISS), which is 254 miles above Earth's surface. Deep-space propulsion is the technology that will take spacecraft to the sun, other planets, and beyond.

Chemical propellants are not optimal for achieving this kind of movement. New propulsion technologies promise faster, more efficient ways to get from point A to point B in space. These include nonchemical propulsion from plasma (electric or ionic), solar sails, and tethers to other moving craft.[5] Nuclear-powered propulsion is an option for in-space and deep-space travel, as well as are lasers and fusion power. Deep-space propulsion may also take the form of "propellentless" technologies like solar electric (SEP).

The propulsion technology industry is one that may place some or all of its operations in space, rather than on Earth. Now, obviously, this is all occurring on our home planet. In the future, and perhaps not too far off in the future, we may see propulsion system manufacturing taking place on the lunar surface or in orbit. This should make sense, given that the systems are not designed for "escape" from Earth's gravity. Indeed, some of the projected lunar industry activity is envisioned as being in support of propulsion technologies.

Space Tourism

People will spend money to travel in space. To date, this is what one might call a "rich man's hobby," given the cost. But, the idea is capturing the public's imagination, especially with publicity stunts like Jeff Bezos' Blue Origin rocket taking William Shatner, Captain Kirk himself, on a suborbital flight in 2021.

Shatner's remark about the experience, "All I saw was death,"[6] might have put a damper on space tourism as a concept, but this has not been the case at all. The space tourism business grossed $695 million in 2022.[7] The industry is projected to grow at a 40% rate in the next seven years as tourists make their way from suborbital to orbital flights.

Space Robotics

Advances in space travel aside, putting people into orbit is still the most expensive way to get things done in space. Robots are far more cost effective, and they have many advantages over human beings. They can persist in the extreme conditions of outer space for years. They can perform all sorts of technical tasks that are beyond human capabilities, and so forth.

Space robotics is already a significant business, with $3.89 billion in industry revenue in 2019 and a projected growth rate of 5.7% through 2030.[8] Current activities run the gamut from repair of satellites to servicing the ISS. Likely avenues of growth include in-space manufacturing and preparing the lunar surface for long-term human settlement.

Imagery Services

Taking pictures of Earth's surface was one of the main raisons d'etre for the development of satellites. The missions were for intelligence, but scientific uses followed soon after. Today, satellite imaging is a $3.7 billion market, growing at a rate of 11% a year.[9]

Satellites in LEO are best for capturing high-resolution images. Customers comprise insurance companies, which use satellite images to detect fraudulent claims; financial firms (which assess investments such as real estate by examining satellite photos of parking lots); and agribusiness, to name a few. Satellites can even pick up things that are hiding in plain sight, such as toxic waste dumps.

Space-Based Internet

It is possible for computers to connect to the Internet through a satellite connection. This idea evolved from a visionary concept to a money reality quite rapidly. Elon Musk's StarLink service, part of SpaceX, is a prime example. The service now fields thousands of LEO Internet connection satellites that provide connectivity to millions of devices on the ground. StarLink is proving to be an important factor in Ukraine's defense against Russia, which disabled much of its terrestrial Internet connections.

Space-based Internet has several advantages over its terrestrial counterparts, which are mostly provided by large telecom and cable TV companies. There's little infrastructure to build on the ground. A small satellite antenna is all that's needed. Putting the satellites into orbit is no minor project, but it's happening. The global satellite Internet market is now around $2.9 billion, with a projected income of more than $18 billion by 2030—a growth rate of 20% per year.[10]

Space Computing Systems

Space vehicles and other equipment, such as satellites and robots, rely on computers to function. If you don't recall, it's worth noting that the Apollo missions were a critical catalyst in the creation of solid-state computers—arguably jumpstarting the modern IT industry. Originally, however, most of the computing power used for space activities remained on the ground. It was too complicated and costly to put large computers into space.

An imaging satellite, for example, might take pictures from space, but beam them back to computers on Earth for analysis. The problem with this approach is that it's way too slow for many modern applications. In a military situation, for instance, instant image analysis could be essential for commanders who need to make decisions in the moment. Waiting for hours for data to move over radio links to terrestrial data centers could put lives in danger on the battlefield.

The solution has been to put more and more computing and data storage onto spacecraft. That way, the data analytics can occur in space. Hewlett Packard Enterprise (HPE) is one of several major technology companies that are pioneering this new industry. They have deployed high-performance computing (HPC) into orbital data centers. This alleviates a lot of network traffic and speeds up whatever process the spacecraft is tasked with performing. It's early days in this category, but an industry is developing around this technology model.

Government/Military/Intelligence

While private industry is the major growth category in space, traditional government, military, and intelligence payloads are still a big part of the picture. Even SpaceX, touted as the corporate future of space, earns hundreds of millions of dollars launching U.S. military satellites into orbit, among other services it provides to the U.S. government.[11] The Department of Defense (DoD) also relies on StarLink for its liaisons with the Ukrainian military—a situation that is provoking controversy as Pentagon officials start to understand their level of reliance on a private company, and its CEO, for a strategic relationship.

The Space Industry Future

The longer term presents some fascinating and potentially lucrative prizes in space. The contours of major futuristic space industries are now fairly well developed, but the time horizon is unclear, and their feasibility remains an open question. In each case, the markets and enabling technologies are unproven or nonexistent. Yet, the potential is easy to see. Some of the ideas being floated today will not work. Others will, and when they do, we'll shift from incremental evolutions in space industry to a revolution. The results will be immense and transformational for life on Earth as well as in space.

The financial sector and governments are taking note and taking action. The Japanese, for instance, announced a $6.6 billion, 10-year funding program in 2023 for the Japan Aerospace Exploration Agency. Funds will go to researching space exploration and funding innovations in the private sector, such as businesses that can compete with SpaceX.[12] The United States has comparable funding programs in effect, and more in the works.

Space Cargo Transport

Getting equipment and raw materials to space and back will require a new generation of space transport vehicles. To this end, the European Space Agency launched a competition to develop a space cargo return service. The goal is to develop a commercial vehicle that can transport supplies to the ISS by 2028.[13] The program is itself interesting, but it is also notable for what it represents: the commitment of money to space transport reveals a belief, in key space policy circles, that space industry is real and that cargo vehicles will be needed to make it happen.

In-Space Manufacturing

Today, nearly every item used in space exploration and commerce is manufactured on Earth and flown into orbit, at great cost. This is about to change. In-space manufacturing is an emerging category of space industry.

The industry, as currently envisioned, has two lines of business. The first is the making of products in space for use in space. This is already happening, in experimental form, with companies like Made in Space using 3D printing on the ISS to create tools. The company was awarded a $74 million contract from NASA to do 3D printing of metals beams for use on NASA spacecraft.[14]

The other in-space manufacturing track involves making things in space that can be produced only in a zero-gravity environment—for sale back on Earth. Fiber-optic cable is one such product. Research has shown that manufacturing fiber-optic cable in zero gravity results in a higher quality end product. Plans are in the works to set up fiber-optic cable manufacturing in space. Certain pharmaceutical products can similarly benefit from zero-gravity manufacturing.

The economics of such industrial operations remain an open question. As of right now, the costs are too high for zero-gravity manufacturing to make sense. However, the existence of sizable markets for the output of in-space manufacturing augurs well for the sector's success. As space transport costs decline—and they will—these industries may soon launch, in more ways than one.

Lunar Industries

The moon, now accessible to no fewer than four earthly nations, is the object of much space industry fantasizing. Overhyped visioneering aside, lunar industries do look like a viable proposition for many reasons. The moon is loaded with valuable resources, especially those needed for space travel and commercialization. The moon is also likely to become the staging point for said space travel and commercialization, so lunar colonization seems inevitable, even if the path the process follows turns out to be different from what we imagine.

First, the moon contains many of the materials needed for long-term human habitation. There is ample water, as well as metals like aluminum, which will enable people to build structures, generate breathable air, and engage in agriculture on the moon. There's unlimited solar power, along with enough native silicon to make lunar manufacturing of solar panels a reality. Hydrogen, useful for propulsion, exists in abundance. As a result, it seems probable that humanity will establish a permanent settlement on the moon at some point in the future, perhaps sooner than we think.

Commercial ventures on the moon will most likely arise from the potential to produce propulsion fuels and manufacture products needed for space travel on the lunar surface. Being able to manufacture rocket engines and the like, without having to haul the raw materials out of Earth's gravitational field, presents a great business opportunity.

Lunar mining is the focus of much business speculation, too. The moon contains large quantities of rare-earth elements, such as scandium and yttrium, which are essential for electronics and electric vehicles.[15] Currently, China is home to 95% of Earth's reserves of these elements, which is not a great situation for the United States and European Union, among others, who don't like depending on China for these economically critical commodities. Getting rare-earth elements from the moon may be costly, but it's the sort of industry that strategic concerns power forward regardless of finances.

The moon is also home to helium-3, an isotope of helium that can power nuclear fusion energy generation both on Earth and on the moon. This is somewhat speculative, because there are currently no nuclear fusion reactors using helium-3, but scientists and certain industrial players see the potential for it.

Success in lunar industries will depend on the maturity and success of related space industries. Without advanced space robotics, for example, there won't be much lunar economic activity. The same goes for low-cost space transport systems.

How big is the lunar economy? It's hard to say, but the moon itself could be home to hundreds of billions of dollars in commercial activity. The moon, acting as the main launch pad for space travel and freight transport, might also be a key component of a far larger space economy.

Orbital Energy Industries

In case you haven't noticed, humanity is in the midst of a perpetual, double-edged energy crisis. We're constantly searching for new sources of energy as we deplete limited resources of fossil fuels—which appear to be rendering the planet uninhabitable. The more we burn, the more we'll drown. We could ditch our cars, switch off all the lights, and go back to the stone age, but that's probably not happening. Space offers a tantalizing solution to this most foundational of civilization's problems.

Solar power works exceedingly well in space. The sun's rays do not have to travel through the atmosphere, which blunts its energy potential. Solar panels in space generate far more electricity than their equivalent on Earth. What if we could somehow get all that energy back to Earth? It would make things a lot better on the planet's surface. We'd have a lot of juice, with none of the environmental impacts.

There is a way to do just this, at least in theory. An orbiting array of solar panels could beam energy back to a specially designed receptor on Earth using microwaves. If such a system could be put into place, we would have an endless source of clean electricity.

It's a fascinating idea, one that could easily turn into a multitrillion-dollar business. The United States currently spends $424 billion a year on electricity.[16] The revenue side of this market is clearly evident. Using present value, that translates into an industry worth more than $5 trillion, and that's just America.

Will someone step up and invest in orbital energy with that kind of valuation in the offing? Today, the answer is a very confident "no." The economics simply don't pencil out at this point. The costs of launching such an array and building the ground receptors would result in electricity costs, on a per-kilowatt basis, that are far higher than those from coal- or gas-fired power stations. One analysis holds that it will cost $194 per watt to generate power in space, versus the $3 it takes on Earth.[17] Power is a commodity business, so low cost always wins.

Thus, space energy will have to wait for some cost breakthroughs or engineering innovations that make electricity from space a lot cheaper. Alternatively, if environmental issues become serious enough, clean electricity from space may become an attractive proposition even if it costs more. At some point, however, it seems likely that we will have electricity from space, and it will be a big business.

Asteroid Mining

The idea that we can send robotic mining machines to the asteroid belt, located between Mars and Jupiter, and have them dig up valuable metals and bring them back to Earth, or the moon, is one of the most ambitious and intriguing ideas in industrial history. The basic premise is viable and well understood: there are hundreds of thousands of asteroids, many of them containing millions of tons of valuable metals like gold and platinum, as well as other materials, such as iron and nickel, that are useful for space—along with massive amounts of water.

Trillions of dollars' worth of stuff is there for the taking. All we have to do is get there and have the equipment to extract it and deliver it where it's needed. And, that, unfortunately, is where reality is intruding on this compelling space industry dream. The costs of building the equipment, getting it to the asteroid belt, overseeing it, and getting the materials to where they're needed are simply prohibitive at this point.

Additionally, the quantities of substances being eyed for asteroidal extraction have the potential to flood terrestrial markets. Gold now trades at more than $1,100 an ounce because gold is rare and difficult to mine on Earth. If a spacecraft drops off 100 tons of gold from an asteroid every month, the price of gold will plummet, voiding the financial justification for the mining mission in the process.

Still, asteroid mining may be the best and arguably the only way to get the materials needed to build spacecraft in space, a process that is considered essential for interplanetary travel and commerce. So, we may yet see a realization of the idea. The question is when. Of all the futuristic space industry concepts, asteroid mining is probably the furthest out in time. We may see it happen in a century, not a decade.

A Coming Great Power Struggle in Space?

As we escape the gravitational pull, so to speak, of the present and start to look further into the future, a number of intriguing potentialities start to emerge. For starters, to see the space industry as a disembodied collection of corporate activities is to miss a critical bigger picture. Corporations, even multinationals, operate inside nation-states. Their gross revenues contribute to their home nation-state's gross domestic product (GDP). Their intellectual property (IP) accrues to the intellectual capital and wealth of the nation-state.

Nation-states compete with one another for markets and growth. There's a link between the economic fortunes of a nation-state and the stability of its politics. Strong business results usually translate into prosperity and entrenchment of political power. The opposite is also true. Downturns fuel unrest. As a result, nation-states compete vigorously for economic growth and domination over rivals. This was true in the eighteenth and nineteenth centuries, and it's true today. It will also be the reality in the space economy.

For some observers, a geopolitical competition for the wealth of space is likely, if not inevitable. Peter Garretson, a former U.S. Air Force officer and senior fellow in defense studies at the American Foreign Policy Council, and Namrata Goswami, an independent scholar, have laid out this argument well in their book *Scramble for the Skies: The Great Power Competition to Control the Resources of Outer Space.*

Garretson and Goswami foresee competition between today's great powers, namely, the United States and China, but also Russia, India, and others, to add huge chunks of the space economy to their earthly GDPs. To put it in practical terms, they're asking, basically, on which national balance sheet will all these trillion-dollar space enterprises belong? If earthlings write a trillion dollars in checks for orbital energy, where's that money going to get deposited? Will it go into the Bank of China or JP Morgan Chase?

As the world well knows, wars have been fought over a lot less money and prestige than the stakes we're envisioning in space. Just as the British and

Russian empires clashed over control of India in the nineteenth century "Great Game," so too might we have the United States and China engaging in endless intrigue and almost-warfare over control of, say, the moon. Or, we could have all out warfare in space. Stranger things have happened.

Pirates have long plied their trade during such times of geopolitical competition. The European powers who battled it out for control over the Western Hemisphere between 1492 and 1900 certainly used pirates to their respective advantages when they felt it was necessary. So, too, will it likely go in space.

From Space Colonies to. . . Space Nations?

If we take the great power analogy further, we start to glimpse an idea that's either fascinating or frightening, depending on where you stand. The race for resources from Africa and the Americas from the fifteenth to nineteenth centuries led to the formation of colonies. Some were set up in the name of the crowns of England, France, Holland, and so forth. Others were corporate enterprises operating with the permission of those crowns.

Eventually, however, virtually all of those colonies became sovereign states. The United States was one of the first examples of this transition. The most recent have been in Africa. This could easily happen in space.

The current call is for humanity to "colonize space." To cite just one of hundreds of examples of this idea, space.com carried an article in 2016 with the title "SpaceX Mars Missions: How They Plan to Colonize Mars."[18] Here, SpaceX would be like the British East India Company setting up shop in Calcutta…and initiating a colonial takeover of the subcontinent.

If SpaceX sets up a colony on Mars, human history tells us that at some point, the people of that colony will demand their political autonomy from their earthly masters. If this happens, then space pirates will find themselves with many opportunities to engage in plunder at the suggestion of the various forces involved in these space-based colonial liberation struggles.

Robert Heinlein anticipated this potentiality in his 1966 book, *The Moon Is a Harsh Mistress*. In the book, lunar colonists rebel against their Earthbound masters, once again demonstrating that science fiction is way ahead of reality when it comes to space.

The Time Horizon, and What It Means

When will all of this occur? We don't know, of course, but Professor George Sowers of the Colorado School of Mines and one of the world's leading authorities on space industry, offered a perspective. He said, "I hope that the economy in space gets to the point where you do have a criminal element, right? Because that says we've won, we've succeeded."

Of course, some of these predicted events will never take place. Others may look radically different from the way people today expect. Future visions are always about conditions in the present, so it's amusing to look back and see how nineteenth-century men envisioned the twenty-first century. Drawings from that time show the skies above New York City crowded with personal blimps. We're probably making the same mistakes about the future of space commerce.

Twenty-third-century space citizens will have a good laugh at our silly ideas about what their lives will be all about. That said, some of this will happen. The timeline, though, is an open question. Will we get electricity from orbit by 2060, or will it be more like 2260?

From the perspective of confronting space piracy, however, the answer doesn't matter. A century might seem like a long time, but it's a blink of an eye in historical terms. The Treaty of Versailles, which ended World War I, was signed more than 100 years ago, yet that (poorly thought out) treaty is still very much with us—affecting conflicts from the Middle East to Ukraine, Africa, and Asia. NATO was formed 75 years ago, and the alliance remains a critical element in keeping the peace in Europe. So it may be, too, with policies and international agreements affecting space piracy. We should keep this in mind as we clock the development of the potentially massive space economy.

Takeaways

Space is already a big business, closing in on a trillion dollars in annual revenue this decade. The future of space business promises to be exponentially bigger. The time frame and specifics of that growth are hard to pin down, however, and it's probable that our predictions and estimates will seem ridiculous from the perspective of the future. However, growth is coming. Key drivers of space commerce include satellites, satellite servicing, in-space manufacturing, space mining, new propulsion technologies, and space-based energy generation. As such lines of space commerce mature and become lucrative, they will become attractive targets for space pirates.

Notes

1. Ted O'Callahan, "Exploring the Business of Space," Yale Insights, October 25, 2022, https://insights.som.yale.edu/insights/exploring-the-business-of-space.
2. Aerospace, "A Giant Leap for the Space Industry," McKinsey, January 19, 2023, https://www.mckinsey.com/featured-insights/sustainable-inclusive-growth/chart-of-the-day/a-giant-leap-for-the-space-industry.
3. Robert Zubrin, *The Case for Space: How the Revolution in Spaceflight Opens Up a Future of Limitless Possibility*, Prometheus, 2019, 408.
4. "2022 Global Space Economy," Bryce Tech, accessed August 9, 2024. https://brycetech.com.
5. Maria C. V. Salgado, Mischel C. N. Belderrain, and Tessaleno C. Devezas, "Space Propulsion: A Survey Study about Current and Future Technologies," *Journal of Aerospace Technology and Management* 10 (2018): e1118.
6. Jackie Wattles, "William Shatner on Traveling to Space: 'All I Saw was Death'," CNN, October 10, 2022, https://edition.cnn.com/2022/10/10/business/william-shatner-new-book-boldly-go-scn/index.html#:~:text=Shatner%2C%20one%20of%20history's%20most,profound%20experience%20of%20his%20life.&text=When%20I%20got%20up%20to,was%20that%20was%20out%20there.
7. "Space Tourism Market Size, Share & Trends Analysis Report by Type (Orbital, Sub-orbital), by End Use (Government, Commercial), by Region (North America, Europe, APAC, Latin America, MEA), and Segment Forecasts, 2024—2030," Grand View Research, July 2024, https://www.grandviewresearch.com/industry-analysis/space-tourism-market-report.
8. "Space Tourism Market Size."
9. Emergen Research, "Commercial Satellite Imaging Market, By Application (Defense and Security, Urban Planning and Infrastructure, Disaster Management, Energy and Natural, and Other), By Vertical, and By Region Forecast to 2032,"

Marketysers Global Consulting, July 2023, https://www.emergen
research.com/industry-report/commercial-satellite-imaging-
market#:~:text=Market%20Synopsis,factor%20driving%20
market%20revenue%20growth.

10. Shadaab Khan, Pramod Borasi, and Onkar Sumant, "Satellite Internet Market
Size, Share, Competitive Landscape and Trend Analysis Report, by Band Type
and End User: Global Opportunity Analysis and Industry Forecast, 2020-
2030," Allied Market Research, July 2021, https://www.alliedmarket
research.com/satellite-internet-market-A12472
#:~:text=Global%20satellite%20internet%20market%20
size,from%20satellites%20orbiting%20the%20earth.

11. Joseph Choi, "Pentagon Awards $160M in Contracts to Musk's SpaceX," The
Hill, November 3, 2021, https://thehill.com/policy/defense/
542728-pentagon-awards-160m-in-contracts-to-musks-
spacex.

12. Mary W. Roeloffs, "Modern Space Race: Japan Pledges $6.6 Billion for Developing
Space Sector as U.S. and China Plan Historic Missions," *Forbes*, November 21,
2023, https://www.forbes.com/sites/maryroeloffs/2023/11
/21/modern-space-race-japan-pledges-66-billion-
for-developing-space-sector-as-us-and-china-plan-
historic-missions/?sh=668694af61b6v.

13. Andrew Jones, "Europe Wants a Private Cargo Spacecraft by 2028,"
Space, November 7, 2023, https://www.space.com/european-space-
agency-private-cargo-spacecraft-2028.

14. Matthew Weinzierl and Mehak Sarang, "The Commercial Space Age is here,"
Harvard Business Review, February 12, 2021, https://hbr.org/2021/
02/the-commercial-space-age-is-here.

15. "The Lunar Gold Rush: How Moon Mining Could Work," Jet Propulsion
Laboratory, May 29, 2015, https://www.jpl.nasa.gov/infograp
hics/the-lunar-gold-rush-how-moon-mining-could-work.

16. "Revenue of the Electricity Industry in the United States in Selected Years from
1970 to 2022," Statista, September 2023, https://www.statista.com/
statistics/190548/revenue-of-the-us-electric-power-
industry-since-1970.

17. "Space-Based Solar Power: The Future Source of Energy?," Green Match, July
22, 2024, https://www.greenmatch.co.uk/blog/2020/02/space-
based-solar-power.

18. Space.com Staff, "SpaceX Mars Missions: How they Plan to Colonize Mars,"
Space, September 30, 2016, https://www.space.com/34251-spacex-
mars-mission.html.

4

Space Piracy: Overview of a Serious, Looming Threat

AI-generated rendering of a space hotel at sunset
Source: Emiliia/Adobe Stock Photos

Now that we have a sense of piracy's history, its present state, and the potential value of the space economy, let's examine the potential for piracy in space. As we saw in seventeenth-and eighteenth-century piracy, as well today at the Horn of Africa, piracy arises in times and places where economic and geopolitical forces make it profitable and advantageous to both the pirates and those who back them. Until recently, space did not meet these criteria, but this is starting to change.

Piracy, like murder, requires motive, means, and opportunity. From the start of the space age until quite recently, a would-be space pirate lacked all three. Yes, they might have motive, a desire to plunder the treasures of space, but that treasure chest was largely hypothetical for many years. In terms of means, space travel was so expensive that only massive, well-funded entities like the U.S. government could even contemplate it. As for opportunity, the lack of means made that a moot point.

Today, with the costs of space transportation dropping and fresh vulnerabilities in space and on the ground, the would-be space pirate can see the contours of a financially fulfilling adventure taking shape. This chapter explores the contours and likelihood of space piracy.

How Likely Is Space Piracy?

Is this really going to happen? That is the big question. For some, including us, the answer is obvious. Whenever you have economically valuable cargoes and other comparable treasures vulnerable to attack, combined with motivated thieves, you are going to have piracy in some form.

We see this off the coast of Africa. Tankers carrying tens of millions of dollars' worth of oil depart from poorly guarded ports, without much in the way of protection. Men (and maybe a few women), many of whom come from impoverished environments where they have few other prospects to make a livelihood, find these rich targets to be too tempting to pass up.

Will this be the case in space? Admittedly, space is a little more complicated as a milieu for piracy. For pirates off the coast of Africa, all you need is a fast motorboat, a gun, and a safe haven and you're pretty much in business. In space, you need a lot of engineering talent, access to space craft, landing zones, and much more if you want to engage in piracy. As Warren Buffet might say, there's a "fence" around the business. Barriers to entry are high. If you can get past those barriers, though, you're in a lucrative occupation.

These barriers are getting smaller, too. While space travel will never cost the same as a zodiac boat, the price of hijacking space cargoes is dropping precipitously. If the process involves computer hacking, well, that may cost even less than a zodiac boat.

Additionally, the human talent available for space piracy is increasingly available on the criminal marketplace. As space powers like Russia suffer from economic dislocation as the result of sanctions, some of that country's space operations people and scientists may be lured into piracy. And, as the financial rewards of space piracy become evident, insiders at space agencies and related organizations may also become tempted to partake of the booty.

Speculative as it may seem, the matter is receiving serious attention. The U.S. Congressional House Subcommittee on Space held a hearing in 2017 on the subject of private-sector lunar exploration. In that hearing, Representative Ami Bera of California asked Professor George Sowers of the Colorado School of Mines, "This is going to be the wild west. Whoever gets that water first, do they own that water? …how do you create the framework that addresses and creates some certainty for folks that are going to go out there and explore and create that market, but then also doesn't stifle innovation?"

Professor Sowers replied, "I view the space economy as a free market, but the free market has to have constraints and controls. There also has to be a system of rights…. There needs to be the establishment of a framework of, if you will, property rights for space, so that if a company invests in developing a resource that they can reap a return on that investment…. And then, you have also the issue of physical security. You know, if there's wealth in space, eventually there's going to be pirates in space and so you kind of have to have to have a way of ensuring the security of the enterprises that take place out there."[1]

We also ran our thesis past a retired member of the intelligence community, who we'll call "Bob," for his views on the likelihood of space piracy. He was at first caught off guard by the idea, admitting the thought had never occurred to him. He said, "It's like a glass sitting there and is unfilled. You have now poured water into it. And nobody even knew that it was there."

Having spent his career discovering and counteracting threats against the United States, Bob thought space piracy was probable to arise in the future. He compared it to airplane hijackings, which were similarly not given much thought before they became a major problem in the 1970s. He said, "We flew coast to coast, and nobody ever hijacked an aircraft. And then, they started hijacking aircraft, which led to the formation of the TSA, Transportation Security Agency, and a whole bunch of new laws. But now we're dealing with a totally new bear, where no one is protected. No one is prepared."

He added, "Space travel is the wave of the future that is going to come. There have been multiple millions of dollars spent on it already. NASA has granted a contract to explore building launch pads on the moon, presumably for reaching Mars. We are getting into space, little by little. It's just a matter of time. It's not a matter of if. It's when an incidence of space piracy will occur. And currently, we are not set up at all."

He further noted, "Now, the first question, how would an incident of space piracy be received in the intelligence community? Well, this is kind of a dicey problem, but first of all, it would be received with a bunch of 'this is not my bailiwick, you handle it' type of responses. The buck can be passed around pretty quickly. Nobody wants to deal with something like this."

On the contrarian side, Dr. Forrest Meyen, cofounder and chief strategy officer of the space robotics company, Lunar Outpost, offered a more pragmatic take on the risk. He said, "Piracy in space has to make economic sense for the perpetrators, and right now, that simply doesn't exist."

What Space Piracy Might Look Like

Like earthly piracy, space piracy will probably exist on a spectrum of illegal activities. On the low-impact end of the spectrum, you will have what might best be called space crime. These would be acts of theft or vandalism in space. As they get more serious and organized—and the value of the thefts grows—you start to verge on piracy.

The boundary between space crime and space piracy is blurry, as it now is on Earth. If a criminal "crew" hijacks a truck carrying valuable freight, is that crime or piracy? One answer may be, "Does it matter?" It does matter to the extent that the patterns of criminal behavior are different, as are the support systems that make acts of piracy possible. And the mitigation of the threat will be different.

Space piracy will differ from run-of-the-mill space crime in that it might include a geopolitical dynamic, such as support relationships between pirate organizations, nation-states, and violent nonstate actors. Space piracy will also likely involve relationships between multiple criminal and noncriminal entities, for example a cartel working with front companies and financial firms.

The way space piracy works will also change over time. Space pirates' activities 20 years from now will be quite different, we imagine, from those of a 100 years hence. Evolution of space piracy aside, we envision it occurring across several broad categories of criminal behavior:

- Disrupting/threatening space operations on planet Earth. This might include extortion, kidnapping of crews, etc.
- Stealing equipment space cargoes—either in space or at the point of reentry
- Disrupting/threatening satellite communications
- Disrupting/threatening orbital serving industries
- Breaching data or space information systems—on Earth or in space
- Disrupting/threatening space communications systems—on Earth or in space
- Stealing or threatening to block satellite operations
- Threatening to sabotage or destroy satellites and other orbital infrastructure
- "Boarding" and taking over in-space habitation areas and industrial facilities, including orbital, lunar, and planetary
- Ransoming space equipment in space, for example, lunar or asteroid mining hardware

Space Warfare vs. Space Piracy

At some point, the spectrum of illegality crosses over from piracy to warfare. As history shows, piracy often occurs as part of a broader military struggle. The same will probably be the case in space.

The funny thing is that although we've been exposed to vast quantities of entertainment featuring wars in space (*Star Wars*, anyone?), the contemplation of actual war in space seems to be a matter of abstraction. Will Russia fight Ukraine on the moon? What would that look like in reality versus the mind of George Lucas?

Serious policy people are certainly focusing on the issue. As but one example, the Center for Strategic and International Studies (CSIS) and its Aerospace Group publish an annual *Space Threat Assessment*. The Defense

Intelligence Agency (DIA) conducts similar research, as does the Director of National Intelligence (DNI). These in-depth analyses focus primarily on the United States' vulnerability to "counterspace weapons" that can impair satellites that help the United States fight wars and defend itself. In particular, they focus on the space war capabilities of China and Russia, as well as Iran, North Korea, and other smaller but potentially dangerous states.

Will space pirates play a role in international conflicts in space? It's impossible to say for sure, but it seems likely. Pirates can move more quickly than a legitimate military. And pirates provide a precious commodity known as deniability. Russia-backed space pirates can hijack Ukraine's space assets, and the Russian government can shrug its shoulders and say, "It wasn't us!" even if no one believes them.

Some might ask, "Does the distinction between space warfare and space piracy matter?" It does, when you think about prevention and response. Defensive alliances for military purposes are different from multilateral cooperation to combat piracy. At least, that's been the case on the world's oceans. In space, it might be different, but chances are, if we see wars in space, and that's almost a certainty at some point, we'll have comparable tactical issues to resolve.

For more insight into this topic, we turned to Tyler Bates, a former U.S. Air Force and Space Force officer who spent time analyzing "irregular warfare" in space in a personal capacity. In his view, space piracy could emerge as a form of irregular warfare "used to advance political or military goals through indirect, deniable, legally vague, or forceful means," as he put it. He elaborated, saying, "Space pirates can also act independently of national sponsors. Space pirates can also be used directly as irregular combatants in a conflict that fight alongside regulars in space. A multipolar world with competing national interests and international norms can be a permissive environment for space piracy."

Privateering in Space?

One intriguing question that comes up in this thought process is "Will there be privateers in space?" The answer is "maybe." While it's tempting to be drawn into a romanticized rewriting of pirate history and misapplying a seventeenth-century notion of privateers in space, this type of activity could very well take place.

We asked Professor Margaret Sankey of the Air University, who has studied the relationships between criminal groups and nonstate actors in geopolitics, and her response was, "Oh, absolutely. There will be the kind of arrangements where a country might, on the down-low, use a nonstate actor that they have a relationship with to damage another country's space assets: just to obscure their own culpability. That's certainly a possibility. It would work the same way as using a proxy group on Earth."

Garretson and Goswami allude to this kind of scenario in *Scramble for the Skies: The Great Power Competition to Control the Resources of Outer Space*. They describe "Piracy" as a "Situational Scenario" in space conflict, wherein, "Individuals from one state might choose to conduct illegal seizure for private gain, which would necessitate intervention by another state against the first state's citizens."[2]

The Four Spaces of Space Piracy

Where will space piracy take place? As we've said, space pirates will likely operate on Earth as well as in space, but the potential reality may be quite a lot more complicated. Traditional piracy relies on constrictions of physical space, such as narrow maritime passages where boats get trapped. Space pirates have access to a far larger playing field. In our view, space pirates will attack their targets and take refuge across no fewer than four layers of space, or "domains," as the military calls them: outer space, terrestrial space, digital space, and false space. The latter refers to the massive fraudulent online ecosystem that masks much criminal activity.

Outer Space

Pirates who can blast off from the Earth's gravitational field will have a very large environment in which to ply their trade. How large? Think about outer space in terms of its spherical size. low earth orbit (LEO) comprises all the space between Earth's surface and an invisible sphere about 1,200 miles out into space. That means space pirates operating in LEO will have an attack zone approximately 315 billion cubic miles in size. For lunar pirates, their sphere of attack will be more than 7 quadrillion (7.13×10^{15}) cubic miles. That's a lot of space to defend.

Terrestrial Space

Space pirates will attack and hide on planet Earth. Indeed, this may be where the bulk of space piracy takes place in the next few decades. And, space pirates may use Earth as a home base for the duration. They may operate exclusively in space at some point, but that's far off in the future.

Earth is a big place, even with today's instrumentation and transportation technologies. Despite the existence of all manner of global surveillance and worldwide military command-and-control networks, space pirates will have no problem making themselves invisible. As the antipiracy efforts at the Horn of Africa show, international forces cannot be everywhere all at once in a 2.5 million square mile area of the ocean. Ashore, the challenge gets exponentially harder.

Hiding in plain sight will also be an attractive option for space pirates. Obstacles to international law enforcement and the touchy boundaries between crime fighting and acts of war will make it possible for space pirates to take up residence in any number of countries and thumb their noses at anyone who dares to come after them. This is what happened in Somalia, and it will probably happen with lunar pirates who return to safe havens on their home planet.

Inner (Digital) Space

Outer space and terrestrial space are analogous to the world's oceans and the European continents of the eighteenth century. At that time, a place like Jamaica might as well have been the lunar surface as far as the average Englishman was concerned. Today, though, we must add a new type of space to the operating zones of space pirates: inner space, also known as the Internet.

When a space pirate collects a ransom for a piece of space equipment, that ransom will almost certainly be paid in crypto currency. It will disappear into the digital ether. The funds may surface elsewhere after being laundered through any number of seemingly legitimate corporate and banking front entities. The pirates themselves will exist in inner space, hiding on the dark web and communicating using encrypted messaging systems.

These are the law-evading tricks currently utilized by the world's hacker gangs and violent nonstate actors. Space pirates, who may emerge from such groups, will use the same techniques. Whoever is tasked with responding to acts of space piracy and bringing the guilty parties to justice will have to be prepared to deal with this digital labyrinth.

False Space

We're going to take the idea of inner, digital space one step further. With the rapid emergence of generative artificial intelligence (GenAI), it is becoming possible for criminals to adopt false identities at a remarkable scale. AI software can generate millions of believable but entirely fake and synthetic identities. Finding space pirates will become a challenge akin to the way one financial investigator characterized the search for a roque commodity trader: it wasn't so much finding a needle in a haystack; it was like finding a needle in a pile of needles.

Who Will Be the Space Pirates?

Like the pirates of the yore, space pirates will likely come from diverse backgrounds that will reflect the twenty-first century in terms of the types of individuals and the skills needed to succeed in space piracy. And, as was the case in earlier times, space piracy will spawn an interdependent, mutually beneficial ecosystem of financial and quasi-governmental partners.

Space pirates will probably emerge from the following sectors, all united in their desire for plunder:

Ex-military personnel: Just as old-time pirates, buccaneers, and privateers had experience in manning ships and military tactics, some space pirates will probably come from the U.S. Space Force, U.S. Air Force, various U.S. missile commands, DARPA, DIA, CIA, NRO, National Geospatial, and other intelligence agencies. These ex-military pirates will not just be from the United States but may come from space powers like the United Kingdom, Germany, Russia, Israel, India, and China. The Los Zetas in Mexico also offer an example of an elite military unit that turned into a criminal organization. (This should not be taken as an indictment of the honor or loyalty of the vast majority of military personnel. As history has shown, however, there are times when former members of the military can be lured into criminal activity. Thinking about space piracy means reckoning with some uncomfortable potentialities.)

Ex-space agency personnel: It is expected that a major source of talent for space pirates will be the world's major space agencies, like NASA; the European Space Agency (ESA); Russia's Roscosmos; the space agency of Japan, JAXA; and other major space agencies from the United Kingdom, Germany, Israel, China, and India.

Corrupt insiders: This is an uncomfortable topic for some, because people don't like to imagine that employees, law enforcement personnel, or members of the military could take part in criminal activities. However, unfortunately, it happens all the time, and space will be no exception. People who have access to confidential information about space assets can be bribed, as can people who are responsible for safe transport of valuable space cargos, and on and on. The potential is endless. With the Internet and cryptocurrencies, such people can hide in plain sight.

Commercial space entities: Personnel from commercial space companies will be prime recruitment targets. These individuals from companies will have unique skill sets. Talent from space startups who are involved in artificial intelligence, machine vision, coatings, software, cyber-robotics, and launch are just a few who will be valuable assets to any space piracy.

Finance/investments: The plunder from space piracy will be large and ongoing. Thus, the pirates will need sophisticated personnel to handle the proceeds, divide it among themselves, hide it from tax and investigative authorities, and invest it. This will require experienced commercial bankers, investment bankers, and experts in commodity brokering, crypto, and blockchain.

Criminal cartels: A major difference between piracy of the past and space piracy will be the startup costs. It is an activity that will be out of reach for most criminal groups. One type of organization, however, has the financial resources and criminal expertise from finance to management to become a player in space piracy. That type of organization is the major criminal cartel, such as those that operate in Mexico. They are not just criminal enterprises but rather close to clandestine state actors. Their power, which provides them access to valuable state resources in addition to capital, will be invaluable to space pirates, who eventually will need a base camp to hide in and operate from. More on this in Chapter 7.

Other sources of pirate talent: To manage what we believe will be a very vast network of criminal enterprises will entail a substantial array of sophisticated talent. It is our contention that the interlocking ongoing pirate entities will have a small permanent band of pirates working full-time, planning the next operations while relying on concealed groups of pirates who will stay in place in key institutions and companies. They will be either infiltrated into the desired organizations or recruited via large bribes.

These moles will provide key intelligence and other operational support activities on an as-needed basis to the pirate bands. Think of how the cartels have penetrated all the significant institutions: police, military, judiciary, prisons, and intelligence agencies. Space pirates will have an analogous network of talent in place in all critical nodes of the space economy.

The motivations of space pirates will not differ much from those of pirates in previous centuries. They will be individuals who are disgruntled with their current career prospects and seeking payback and plunder. There will be adventurers, and there will be individuals who will see more career opportunities in a space pirate organization.

Financing and Equipping Space Pirates

Space is expensive, arguably one of the most capital-intensive businesses in the world. How will space pirates, buccaneers, and privateers obtain capital to engage in space piracy and plunder? In some cases, they may have their own ill-gotten money to invest in space piracy projects. However, they may also be able to access capital through seemingly conventional sources like investment banks and venture capital funds—perhaps by posing as legitimate business ventures.

In terms of outfitting space pirates, the most probable resource will be a rogue spacefaring nation. Russia is a good candidate for this role, but it is not the only one. Russia's space program—which inherited the Soviet space program—has many impressive achievements to its name, including its participation in the International Space Station and providing engines for U.S. rockets. However, the program had serious funding, management, and operational problems that have now been exacerbated by U.S./European sanctions against Russia due to its war against Ukraine.

As the Russian space industry continues to decline because of the sanctions, it is very likely that if the cartels and space pirates and buccaneers approached Russia with an offer to become a partner in their space piracy activities, it would be very difficult for Russia to refuse. While the financial rewards would be attractive to Russia, of immediate and critical importance would be that a partnership with the cartel and space pirates and buccaneers would allow Russia to keep the doors of its space agency and industrial base open. This might be especially true given its space program's very sensitive military aspects (e.g. missile development). There is no distinction between civilian and military uses of space in Russia's space program.

A Russian deal with the cartels, space pirates, and buccaneers would create one of the most formidable space programs in the world. Russia can bring the following extremely high-value assets to this partnership:

- **Operational expertise:** This will include but not be limited to launch capabilities, ground station tracking, and a bench of experienced cosmonauts who can be used for attacking space stations, spaceships, and planetary habitations.
- **Manufacturing of space platforms:** This will include expertise in the manufacturing of rockets and rocket engines, and space station construction and management.
- **Space communications infrastructure manufacturing:** This will include encrypted space communication devices/software, satellites, optics, and sensors.
- **Science/design:** This will provide access to hard science for advanced rocketry, planetary habitation, and space biology; new coatings, materials, and alloys for space transportation and planetary habitation; and propulsion systems ranging from chemical to electric and nuclear.
  - Access to intelligence via SVR and GRU to aid in planning pirate and buccaneer attacks.
  - Access to military weapons (e.g. lasers, ground-based anti-satellite systems, hypersonic missiles).
  - Access to the training of the next generation of space pirates and buccaneers via space/military professionals.
  - Access to financial institutions off the Western grid.

This is not an entirely speculative idea. As of right now, Russia is threatening to attack the West's satellites in retaliation for its support of Ukraine. With Russia at war with the West, it would be very tempting for Russia to engage in space privateering to protect its interests via taking offensive actions against the space assets of its adversaries. With a relationship with the cartels, space pirates, and buccaneers, Russia would have an immediate "off-the-shelf" group ready to execute its wishes.

Takeaways

The probability for space piracy to take off as a criminal activity seems high. As the value of space cargoes and equipment rises, thieves will no doubt be attracted to them. Pirate activities in space, or on Earth-bound parts of the space industry, will likely involve seizing goods and equipment, as well as people, and holding them for ransom. These acts may be part of bigger geopolitical conflicts, as was the case in the seventeenth and eighteenth centuries. Piracy provides deniability, which is attractive to nation-state actors. Space piracy will be costly, though, and sources of funding are unclear. However, given the potential "return on investment," so to speak, capital may make itself available, possibly via mainstream business ventures that get coopted into piracy by rogue insiders. A rogue spacefaring state would have to be part of the story, making its technology and know-how available for acts of space piracy.

Notes

1. U.S. Congress, House of Representative, Committee on Homeland Security, *Challenges of Recruiting and Retaining a Cybersecurity Work Force*, 115th cong., H. Rep.115–26, `https://www.govinfo.gov/app/details/CHRG-115hhrg28415/CHRG-115hhrg28415`.
2. Namrata Goswami and Peter A. Garretson, *Scramble for the Skies: The Great Power Competition to Control the Resources of Outer Space*, Lexington Books, 2020, 307.

5

Space Hacking: Current Realities and Future Lessons for Space Piracy

AI-generated rendering of space hacker
Source: Shutterstock AI/Shutterstock

In August 2023, the National Counterintelligence and Security Center (NCSC), which is part of the U.S. Office of the Director of National Intelligence (DNI), issued a bulletin that warned of cyber threats from foreign intelligence services on America's growing space industry. The bulletin also highlighted the risks to space commerce from shell companies and other forms of espionage.[1]

For people who pay attention to the world of cybersecurity, the bulletin probably registered a "duh!" but the issue is nonetheless serious and deserving of attention. It landed two years after the notorious SolarWinds "supply chain" attack resulted in breaches of systems at NASA and the Treasury Department, and a year after hackers started trying to disrupt StarLink satellites providing Internet service to the Ukrainian military. There had been many other instances of hackers targeting satellites in previous years, as well.

Computer systems are foundational to space commerce. Space travel, transport, colonization, and business depend on computer hardware, software, databases, and networks. Given the sorry state of cybersecurity, this reliance should be a cause for concern, if not outright alarm.

Keep in mind, further, that space cyber assets may be more significant than they appear on the surface. StarLink is an Internet service provider based in space. That makes it important for any number of industries and governments, but there's more to it. According to Naeem Altaf, a distinguished engineer and chief technology officer (CTO) of Space Tech at IBM, StarLink has the potential to become the digital backbone of the entire space ecosystem. "StarLink is the backhaul segment of the Internet in space," he explained. He further added, we need to make sure the networks in this hybrid configuration for space tech should be quantum safe. It means they are designed to be secure against attacks from quantum computers using quantum-resistant cryptographic techniques.

And, as recent events show, hackers are already attacking vulnerable systems in space, but the risks are actually far more extensive: space industries and related government and military space activities connect to, and are to some degree dependent on, a vast universe of digital assets spread out across the world. These include corporate and government networks, information systems, and much more. A breach of any one part of this massive, interconnected system could cause problems for space operations.

Space pirates, who are arguably already among us in the guise of hacking gangs, are keenly aware of this—and are likely to continue exploiting widespread, systemic security weaknesses for their own financial gain. At some point, today's space hacking will evolve into digital space piracy.

Additionally, hackers will almost certainly comprise a key support system for other modes of space piracy, such as kinetic attacks on space assets. This chapter explores cyber risks in space and their relevance to the evolution of space piracy.

The Current State of Cyber Insecurity

We want to touch briefly on the current state of cybersecurity, though the word "insecurity" might be more appropriate. Numerous excellent books cover this subject in great depth, and you are probably familiar with some of this already. So, without going too deeply into issues covered more extensively elsewhere, our goal is to highlight the major areas of cyber risk as they relate to space commerce and the potential for space piracy.

Software Risks

Software makes computers do what they do. Some software is right in front of you, like a web browser. Other software programs, like operating systems and "firmware," are buried deep down in the "tech stack," where most of us have no idea what's going on. All of it is vulnerable to attack, however, and any software compromise can affect a computer system's ability to function.

Malicious actors like to target software because software handles data and operational processes. Want to steal data? Hacking applications that handle data is a good way to go. Want to stop a process, like launching a rocket, dead in its tracks? Hack the software.

In some cases, it's possible to hack software without penetrating a network and compromising a specific "endpoint device," such as a server or laptop. Success depends on a reality that's not well understood outside the software industry: most software today is created by assembling code blocks from preexisting open-source libraries. In a "supply chain" attack, hackers implant malicious code into those code libraries. The software thus ships with a virus (malware) already embedded in it, ready to open back doors to data and worse.

The 2020 SolarWinds attack is one of the most recent and disastrous examples. Thousands of companies and government agencies used SolarWinds for system management. The software was thus connected to almost every major system at client sites.[2] By compromising SolarWinds, the attackers got easy access to a huge range of critical systems. It wasn't so much the fox guarding the hen house, so to speak, but rather like the fox designing and building the hen house.

Satellites and space vehicles are platforms for software. One could argue that they are network "endpoints" just like smartphones. They're just orbiting Earth. As such, they are vulnerable to attacks on the software they require to function.

Data Risks

Data is frequently the target of a cyberattack. The hacker's goal is to steal, erase, or deform data, with the driving purpose being either making money or disrupting operations—or both. We hear of massive data breaches nearly every day right now, unfortunately. Major health insurers, credit rating agencies, huge government agencies, and Fortune 500 corporations have all suffered data breaches.

Ransomware, a form of cyberattack that involves encrypting the victim's data until he pays a ransom for decryption, is one of the most damaging, common, and lucrative threats to data today. It is also, arguably, similar to piracy. Like Somali pirates who board a tanker and demand a ransom for its release, so too do Russian hacker gangs encrypt corporate data and demand a ransom to give it back to its rightful owners.

Ransomware is about more than just making money by hacking, however. A ransomware attack can badly impair the victim's ability to make money. The target may go out of business, or cease to function—with deadly results. Indeed, several people have died in recent years after hospitals where they were receiving treatment became unable to deliver healthcare due to ransomware attacks.

To date, ransomware has not been a big issue in space, mostly because space systems are difficult to hack, and most of them do not store a lot of data that can be encrypted. That is starting to change, however, as more space systems carry "off-the-shelf" computers and store data to avoid transmitting it back to Earth for processing.

Network Risks

In many, if not most, cyberattacks, the attacker must gain access to the target's network. Once inside, the attacker can move laterally and attempt to access software programs, databases, user identity stores, and the like. But, getting in is a critical step.

For this reason, networks generally have robust defenses. These include firewalls and intrusion detection systems (IDSs). Organizations also deploy various forms of user authentication, such as SMS one-time passwords (OTPs) to verify if a user is actually who they say they are. Still, network defenses can have weak spots, and determined attackers tend to find their way inside.

As of today, many space assets are not connected to terrestrial data networks or the public Internet. They are "air-gapped." Large-scale space assets, like the ISS, have their own internal data networks. However, as cybersecurity experts can tell you, air gaps are not always so effective. Computers and other devices are often connected to the Internet by mistake, leaving them vulnerable to attack. As space commerce expands, Internet connectivity from space will likely grow. And space assets will be connected to one another through space-based networks. All are vulnerable to penetration by hackers.

Hardware Risks

Hackers can compromise computer hardware, a process that can result in deeply damaging and hard-to-detect cyberattacks. Most cyber defense systems and processes are geared toward stopping penetration of networks or takeovers of user accounts. The hardware itself tends to be an afterthought, partly because hardware attacks are difficult and beyond the reach of the average hacker. However, when attackers are able to take over the hardware, they can wreak havoc on the systems it supports. Working at the hardware level, attackers can feast on sensitive data and corrupt any number of system processes.

Hardware security is a major factor in the growing Internet of Things (IoT), which involves the deployment of billions of digital sensors and comparable devices to measure things like the weather and industrial production. The IoT is a security nightmare, as cheap devices, often with little to no innate security, are gathering and storing data needed for all sorts of vital processes in business and society.

Satellites are analogous to IoT devices. They are autonomous, remote computers that take in data, perhaps processing it a little and then forwarding it either to a ground-based data receptor or to another space-based computing system. As such, they are vulnerable to hardware attacks.

This is not a hypothetical risk, with recent revelations about Chinese manufacturers installing backdoors on encryption chips used by the U.S. military,[3] and other warnings about the Department of Defense's (DoD) worrisome dependence on China for the electronics it needs to fight wars.[4]

Denial-of-Service (DoS) Attacks

Cyberattackers often take over numerous computer systems, allowing them to function while secretly operating them for nefarious purposes. When they amass enough computers in this sort of so-called bot net, they can send messages to a target system and flood it with so many requests for processing that it shuts down. This is known as a *distributed denial-of-service* (DDoS) attack.

DDoS attacks are common and potentially serious. For example, in September 2023, Akamai, whose massive content distribution network (CDN) is ideally situated to observe and react to cyberattacks, observed and thwarted a massive DDoS attack targeting a large American financial institution.[5]

As Akamai explained, there is usually only a small amount of legitimate traffic coming to this company's site from within the United States. However, in just two minutes, the target was on the receiving end of 633.7 gigabits of traffic per second (Gbps) and 55.1 million packets per second (Mpps) from all over the world. Sources included Bulgaria, Brazil, China, India, Thailand, Russia, Ukraine, Vietnam, and Japan.

DDoS attacks can be a nuisance, but if they target critical digital assets, such as a satellite ground station or the systems controlling a space vehicle, the results could be catastrophic. A space pirate could mount a DDoS attack against a space asset and discontinue the attack only upon payment of a ransom.

OT vs. IT

The computers most of us use in day-to-day work and corporate operations are known as "information systems" or "information technology" (IT). These are mostly familiar machines, such as those powered by Intel X86 chips and running Windows or Linux operating systems. However, there is another mode of computer system that is essential in industrial operations, power generation, and other critical infrastructure. It's known as operational technology (OT), which differs from IT in terms of its operating systems, data management processes, and more.

OT systems, such as supervisory control and data acquisition (SCADA) systems, are often older than IT systems at use in the same organization. An industrial company might have the latest version of Windows running on new laptops. The same company might easily have SCADA systems that are 20 to 30 years old running their factory equipment, elevators, and building management systems. The electrical power they use comes from a generation station running SCADA systems.

Due to their age, OT systems are highly vulnerable to attack. Many of these systems were designed before today's cyber threats were a serious concern. Defending OT is hard, and replacing OT systems is a complex, costly proposition.

The extended industrial ecosystem of the space economy is rich with OT systems. This will be a factor in defending the space economy from hackers who target OT.

The Size and Scope of the Problem

Defending computer systems, data, and networks from malicious actors is a big business. Estimates vary, but most analysts estimate the size of the cybersecurity industry at around $200 billion a year.[6] The industry, along with in-house security teams, keeps at least 300,000 people busy full-time. These numbers are only going to go up. Cyber disasters are getting worse, and the field is short of skilled employees.

The Changing Nature of the Attackers

In the popular imagination, hackers are hoodie-clad teenage malcontents who peck away at keyboards in their parents' basements. While there are surely hackers who fit this mold, the more relevant threat is a great deal more serious in nature. Successful hackers today mostly come in the form of organized criminal gangs operating outside the United States in places like Russia, North Korea, and the Balkans. Other times, hackers comprise elements of nation-states' intelligence services.

The Chinese state intelligence apparatus, to name one big offender, has targeted numerous private corporations and government agencies in cyberattacks. For example, in 2018, hackers suspected to be from Chinese intelligence penetrated the data systems of a U.S. Navy contractor and stole secret data about submarine warfare and secret codes.[7] We can only imagine that such attacks will target space assets, if they haven't already.

The armed forces of India, for one, have few doubts about China's intentions to hack satellites. An article in *India Narrative* with the suggestive title "Is China trying to hack into satellite communication as part of asymmetric warfare?" contained claims from Indian intelligence people that "China is now experimenting in space for its non-kinetic warfare strategies. It is believed to be trying to hack into satellite communication of other countries and hijack these satellites."[8]

To be fair to China, this is not a one-sided threat. The United States and others are hacking computer systems in China, as well. China, for its part, has complained to the United Nations that StarLink satellites have interfered with their Tianhe space station.[9] This is not to justify the behavior, though. The risks are still serious.

Additionally, hackers are seldom individual people anymore. In almost every serious case of hacking, the attacker is actually a collection of software robots using automated processes and AI to penetrate as many targets as possible in a short period of time. This fact is relevant when considering how to defend space assets against cyber threats.

Bottom Line: Everything Has Been Hacked

If you talk to experienced cybersecurity people, when they're away from microphones, they will tell you that every significant network and computer system in the world has been hacked—probably more than once. This is just a reality of our age. Data and systems are not safe.

There's no reason to imagine this won't be the case in space. Chances are, the computer systems that support the current space programs and commercial ventures in space have already been thoroughly penetrated as well. This may be confusing, as most of these systems seem to be functioning just fine.

That misses the point, however. Hackers often penetrate systems to install silent implants and hidden back doors they can use later. It is possible that nearly every computer system and network powering space activity today contains ticking cyber time bombs that malicious actors can set off when it suits them.

There Have Been Some Successes in Cyber Defense

It is easy to get discouraged about the sorry state of cybersecurity, with good reason, but there have been successes in cyber defense and reasons to be hopeful about things getting better in the future. Defense in depth, for example, which involves deploying layers of defense akin to multiple physical perimeters on a battlefield, can do a lot to mitigate cyber threats. Innovative engineering is leading to the creation of sophisticated cyber defense technologies, many of which use AI to detect attacks before they cause damage.

Digital targets are also protected by a simple fact of life in the world of hacking: it's difficult, bordering on impossible, to successfully attack multiple targets at the same time. Professor Martin Libicki of the U.S. Naval Academy makes this point forcefully in *Cyberspace in Peace and War*, his important, foundational text on the subject.[10]

If a cyberattack requires two or more targets to be compromised at once, the likelihood of failure rises dramatically. This protects critical infrastructure, such as power generation stations, as well as military targets. It will also prove helpful in defending space systems, which are likely to be heterogeneous and interconnected, from the most serious cyber threats.

How We Got Here

Good news aside, anyone interested in protecting space from hackers should ask, "How did we get here?" How is it that computer systems and networks, on which the entire world arguably depends to function these days, are so insecure? It's an interesting topic, in general, but the issue is critically important for understanding vulnerabilities faced by space assets in the present and future.

Like all difficult questions about big issues, the answers are not simple. There are dozens, if not hundreds, of factors that explain why hackers can run roughshod over so many expensive security platforms and breach sensitive data at will. At its root, however, our current cyber insecurity mess—or disaster, if you will—is the result of a failure of imagination. As computer systems evolved over the years, their creators had trouble imagining the kind of world they were creating—and how their technologies would be open to attack.

It's a long story, one that's been discussed in depth in many venues, so here's a highly truncated version: when computers first came on the scene in the late 1940s and early 1950s, they were large, expensive machines that generally lacked connectivity to anything except a few dedicated terminals. Security was not a big concern because only a few people could even access the machine in the first place.

The situation was analogous to early space exploration. The equipment was big, costly, and limited in functionality. Only the government or major corporations could afford it.

As computers proliferated and became smaller in the 1960s, security was still not a major issue because, again, connectivity was limited, and computers remained the preserve of trained, trusted people. In the late 1960s, however, things started to change. The Department of Defense sponsored the creation of the Advanced Research Projects Agency Network (ARPANET), the first public packet-switched computer network, in 1969.

ARPANET was the technological foundation for the modern Internet. And, despite being created for use by the military, it did not embody much innate security. To be fair to its creators, they were not thinking about billions of people around the world having access to it, so they didn't think too much about security. Internet users were either government employees or university students and staff members. These were people who belonged to trusted communities, so the idea that they would use the network for crime did not apparently cross too many minds.

In the early 1990s, the Internet became public. Anyone with a computer and modem could access it. It gained new layers of technology, such as the World Wide Web, which enabled people to share documents online. Security was still very slender, because no one quite imagined what it would become. A few people, like Jeff Bezos, had an idea of the commercial potential of the Web in the early 1990s, but he was a rare visionary.

Thus, as the Internet exploded in popularity, its security flaws became evident. For example, the standards that defined the Internet allowed people to use IP addresses anywhere in the world with little traceability. It still is quite difficult to determine exactly where someone is located on the Internet. This is a big security risk factor. The industry has been in a desperate scramble to catch up ever since.

The Internet's creators have offered some mea culpas. Vinton G. Cerf, the computer scientist who coinvented the TCP/IP protocol that arguably makes the entire Internet function, said, in 2015, "We didn't focus on how you could wreck this system intentionally. You could argue with hindsight that we should have, but getting this thing to work at all was non-trivial."[11]

In a *Washington Post* article about cyber insecurity, David D. Clark, an MIT scientist who was an early participant in the development of the Internet, similarly reflected on today's massive Internet hacks by saying, "These developments, though perhaps inevitable in hindsight, have shocked many of those whose work brought the network to life, they now say. Even as scientists spent years developing the Internet, few imagined how popular and essential it would become. Fewer still imagined that eventually it would be available for almost anybody to use, or to misuse."[12]

Compounding the innate security weaknesses of the Internet was the trend toward "open architectures" in computers and network technology. Companies could make hardware and software in any way they wanted, as long as it adhered to some basic standards of operation. This was a huge boon to the industry. Indeed, open architecture is arguably why we all have low-cost personal computers at all, but it is a security nightmare. Security is up to each device or software maker, and variations in security cause gaps that attackers exploit.

If you're wondering if these risk factors are present in the development of space technologies, you're thinking about this issue the right way. Space systems are now embedding all manner of insecure IT into the core designs. All the vulnerabilities that affect computer systems and networks in the corporate and government realms are present in space. That's a big problem if you're worried about space crime and space piracy.

The Cyberattack Surface in Space

Cybersecurity professionals refer to parts of a computer system or network that are accessible (at least in principle) to hackers as its "attack surface." With the ability to hack a satellite a proven fact,[13] defining the space cyberattack surface is no longer a theoretical activity. People interested in defending space commerce and other space systems from piracy should consider the following elements of the attack surface:

- Hardware and software that runs satellites and ground stations
- Hardware and software that is core to businesses and government entities that support space travel and transportation
- Hardware and software that runs life support and related capabilities on space vehicles and space habitats
- Space-related networks
- Any computer system that supports communications with space vehicles, habitats, or their terrestrial components
- The software and hardware supply chains of makers of space equipment

A 2013 article in *InfoSec*, a respected cybersecurity publication, offered its conception of major threats against satellites,[14] including the following:

- Tracking of satellite data and software
- Listening (eavesdropping) in on satellite communications
- Interacting with satellites without authorization
- Taking over a satellite's functions
- Breaking obsolete satellite technologies such as the X.25 data transmission protocol
- Jamming radio frequencies used by satellites
- Mispositioning/control of satellites by spoofing transponders or hacking command-and-control (C2) systems
- "Grilling," which involves activating all of a satellite's solar panels when exposed to sun, which results in overcharging the craft's energy system
- Causing collisions between satellites

These lists are most likely incomplete, but they get at the size and depth of vulnerability-facing entities that want to be active in space commerce and exploration. As of now, all of these areas of the space cyberattack surface are poorly or at least irregularly defended. The cyber risks to space entities are significant as a result.

Professor Libicki offered an example of how risks in space could play out. As he put it, "Every satellite has a command and control system that relies on a ground station and software on the satellite. A bad actor could take control of the satellite and order it to self-destruct by using its remaining fuel to push itself out of its designated/proper/designed orbit."

Such an attack is possible, according to Libicki, because of trade-offs that get made between those who defend the satellite and those who need access to it. He said, "If you're blocking access to the satellite's communication channel for security reasons, you are probably going to make life difficult for people who need to get data from the satellite, update its systems, and so forth. If there are burdensome permissions processes and encryption technologies at work, there will be complaints, and people might get lazy, such as by sharing encryption keys. From this, you get cyber risk exposure."

According to Mathieu Bailly, VP of space at Cysec, a maker of security software for space systems and other networks, the most urgent need affects commercial satellite operators who are rushing to put their constellations in space. He said, "They want to be the first, but they're working with teams that have little knowledge of security." As he explained, one can graduate as an aerospace engineer but have little training in security. And, he noted, there are few off-the-shelf security products for space assets.

Harrison Caudill, in his paper "Big Risks in Small Satellites," resonated with this viewpoint, commenting that a satellite constellation can be viewed as a "compute cluster" comprising multiple independent systems. He said, "Satellites are, for the most part, computers with peripherals and network connections. The network connection doesn't work the same way, the usage model differs, and the stakes are higher, but the fundamentals are largely unchanged. The differences between space systems and traditional compute clusters must inform any space cybersecurity solution."[15]

Space Hacking Incidents

It is a little odd that entities like the U.S. government are warning of potential hacking of satellites when such attacks have already occurred. The website Space & Cyber Security publishes a list of space cyberattacks. The site lists dozens of attacks, going all the way back to 1977's hijacking of a commercial satellite carrying audio signals.[16]

In another example, in 2007, the Sri Lankan terror group Liberation Tigers of Tamil Eelam (aka the "Tamil Tigers") hacked an Intelsat satellite.

They used the satellite to broadcast radio and television messages. The incident was confirmed by Intelsat, which met with the Sri Lankan ambassador to the United States to attempt to resolve the matter.[17]

Hacking a Satellite: Harder Than It Looks?

A confusing question emerges from all of this concern about the cyber vulnerabilities of space assets. If hacking satellites is easy and hackers can disrupt multimillion-dollar businesses by taking them over—and demanding ransoms to set them free—why is satellite hacking not an everyday occurrence? Satellite hacking is rare. Why?

The short answer is that, as of today, hacking satellites is a lot harder than it looks. For more insight, we turned to Aaron Moore, CTO of QuSecure, the quantum security company. Moore has spent his career in space technology and space cybersecurity, with stints at Raytheon, Northrop Grumman, the NRO, and DARPA.

We asked Moore a basic question: why haven't more satellites been hacked if they are supposedly so vulnerable? For instance, if there are thousands of satellites in orbit, many with unencrypted data and obsolete software, why aren't hundreds of them getting taken down by ransomware? It would seem that there's a lot of money to be made in that kind of attack.

Moore replied that a ransomware attack on a satellite would be a very difficult thing to pull off. "For a number of reasons, it's not really an ideal target for ransomware," he said. "There's not a lot of persistent data that remains on a vehicle itself, so there isn't much to hold hostage."

Like Professor Libicki, Moore felt that a hacker's best bet would be to get into a satellite's C2 system. "They could lock it up, making the satellite useless," he said. "But, there are a lot of barriers to that, too. The waveforms and protocols that are being used to communicate to the vehicle itself and its payloads are usually segmented from the C2 system on the platform. In fact, a lot of satellites have two separate C2 systems: one for the satellite and one for the payload. It makes hacking a lot more complicated to pull off."

He went on, saying, "You've got an executive on the platform, which you can think of as an operating system. They used to be much less robust. They were really only for the functions necessary to communicate directly to the hardware or the satellite platform. Your ability to have functions in an operating system was very limited. It was therefore quite a small attack surface. The instruction sets themselves are custom in a lot of cases, especially for sensitive satellites."

This customization of instruction sets makes satellites hard to hack with "off-the-shelf" hacker tools, which are designed for mass market operating systems and applications. For instance, hackers have developed many tools to take over Linux and Windows servers, but these tools are not readily adaptable to custom instruction sets on satellites. Most satellites are not running Windows or Linux. Rather, they run real-time operating systems (RTOSs). A hacker would have to create specialized tools for an attack, a difficulty that deters a lot of malicious actors from trying to hack satellites.

Regarding data on satellites, Moore said, "The problem with the older satellites, of course, is that they used older modes of encryption. So yes, it's very feasible to get into that. But then you're talking about sophisticated hacking that would require breaking encryption. Now, some satellites don't have any encryption, and that's a problem obviously. But then the data on them is perishable. It's in formats that usually don't allow easily to them to be easily interpreted." [sic]

At the same time, Moore warned, as satellites modernize, they are starting to carry more standard IT assets like X86 servers and commercial database software, which are vulnerable to standard hacking techniques. Not that it would be easy, exactly, as he explained. An attacker still has to establish communication with a satellite, which might require hacking a ground station that is, itself, "air-gapped" from publicly accessible networks. "Anything can be hacked," he said, "but each countermeasure adds to an overall defense that's hard to penetrate."

The use of data diodes is a further obstacle to satellite hacking. A data diode is a hardware appliance with a data transmitter on one end and receiver on the other. As Moore explained, with a data diode, data does not flow in two directions. It flows in one "so you can push data from a secure environment down to a low-security environment or up to a highly secure environment, but there's no communications between the two," Moore pointed out. "That means it's impossible for a ransomware attack to succeed because a malware agent cannot establish bidirectional communi-cations. This is one of the biggest advantages within the satellite architecture."

Does Moore worry about any aspect of satellite cybersecurity? Yes, he is concerned about a supply chain attack. Though, as he admits, the bar is quite high for such an attack, if malicious actors can implant malware into a satellite's code at the development stage, a lot of bad outcomes are possible.

He is also concerned about physical (kinetic) attacks on satellites as well as denial-of-service (DoS) attacks. In his view, that's basically electronic warfare, for example, jamming. "If you look at a satellite signal coming down," he said, "in terms of power, which is regulated in at least in our government, you don't get a lot of power hitting the ground. It doesn't take a lot to jam that signal."

We then asked Moore what he would do if he were a "moustache twirling villain" who wanted to hack a satellite. Whom would he hire to do the deed?

He replied, "I would get people who have built satellite payloads before, people who understand normal satellite office communications and satellite bus communications. I'd get people who were very familiar with vulnerabilities with runtime RTOSs, as well as folks who were very savvy with electronic warfare as delivery mechanisms." The question, then, is whether people with such skills might be tempted to go to work for the bad guys.

Differentiating Space Hacking from Space Piracy

Cyberattacks exist on a spectrum of severity, perpetrated by a variety of malicious actors. Some attackers are well organized, such as those in a criminal gang or affiliated with a nation-state's intelligence services. Others are less organized, such as lone hackers operating on their own. Impacts vary, as well. At the low-impact end of the spectrum, some attacks are simply a nuisance or acts of "hacktivism" meant to publicize a political point of view, for example, digitally defacing the website of a meat packing plant to protest mistreatment of animals.

The following table breaks the organization/impact spectrum into four quadrants. Space piracy through hacking, if it is to occur, will most likely involve highly organized attackers who are intent on having a major impact on their targets. This is different from what is now occurring, for the most part, at sea, where attackers are from small, technologically sophisticated groups. Space piracy hacking will probably be more like seventeenth- and eighteenth-century buccaneers and privateers, who were organized and equipped with the high-tech weapons of their day.

High level of organization, e.g. a criminal gang or coordinated group of hackers Low level of impact	High level of organization High level of disruption E.g. Nation-state sponsored ransomware of major digital entities Large scale fraud Piracy
Low level of organization, e.g. lone attackers Low impact/intent, e.g. nuisance or symbolic "hacktivism"	Low level of organization High level of impact (Less likely to occur)

Level of organization on the attacker side

Impact/intent of disruption →

What is the difference between high-impact space hacking and space piracy? In our view, the differentiating factor is one of intent. If a hacker group wants to disrupt space operations by breaching critical systems, that is not piracy. If they demand a ransom to reactivate those systems, that is starting to look like piracy.

A good question here might be "Does it matter?" From a pure cyber defense perspective, the answer is no. Hacking is hacking, regardless of the nature of the attacker and their intent. However, bigger picture, understanding whether the attacker is engaging in piracy or simple cybercrime, should make a difference when it comes to mitigating risks.

For example, most sophisticated cybersecurity teams tap into threat intelligence data feeds to monitor potential threats to their digital assets. In some cases, they participate in an information sharing and analysis center (ISAC), a nonprofit organization that gathers and disseminates cyber threat information on a centralized basis.

There are ISACs for many industries, such as power utilities, the military, and so forth. If things go as we imagine, there will be an ISAC for space piracy hacking coming along in the next few years. Cyber defenders will want to subscribe to that ISAC if they want to be aware of potential threats from space pirate hackers.

It was in this spirit that the U.S. DoD announced, in September 2023, that it was going to provide more resources and intelligence to the private sector as part of its 2023 cyber strategy. This is probably a wise move because private industry operates nearly 90% of the country's critical networks.

Lt. Gen. Charlie Moore, who was deputy commander of Cyber Command from 2020 to 2022, offered his perspective on the shift in strategy, saying, "DoD's cyber strategy was extremely reactive in nature and led U.S. Cyber Command to really only be prepared to help recover from a cyber event and to develop capabilities that would only be used during war. During those days, I would frustratingly refer to Cyber Command as the 'clean up on Aisle 6' and 'break glass in time of war' command."

Mieke Eoyang, DoD deputy assistant secretary for cyber policy, commented on the new public–private partnership approach by saying, "We owe you actionable intelligence, and you will defend the networks yourselves."[18] This approach is not without controversy. Some argue that the DoD's job is to defend U.S. interests against foreign enemies. If a foreign intelligence service mounts a cyberattack on an American corporation, how is that different from that foreign country firing a missile at an American target? The DoD is clearly on task to defend against missile attacks. Why not cyber?

The official answers to these questions are not relevant. The reality is that cyberspace is simply too large and complex a domain for the DoD to defend with any coherence. Everyone knows that, so they seek public–private partnerships to mitigate the risks. It's probably the best we can do at this point, but it doesn't augur well for defense of space assets against cyberattacks. It is likely that private corporations will be largely on their own when it comes to cyber defense in space.

Existing Cybersecurity Standards and Practices in Space

It would be unfair and inaccurate to say that no one is taking cyberthreats in space seriously. As with cybersecurity on Earth, a lot of smart people are focusing on the risks and offering the best solutions they have available. In 2023, a cybersecurity firm called SpiderOak, for example, announced a collaboration with the Space Development Agency, part of the U.S. Air Force, to develop space cybersecurity solutions.[19]

Additionally, most corporate and government cybersecurity programs are based on guidance from the National Institute of Standards and Technology (NIST) cybersecurity frameworks. This is because NIST is the usual "go to" resource when the government establishes a policy around cybersecurity, such as in a presidential executive order.

NIST has published numerous frameworks for security in different industries and government settings. Their Cyber Security Framework (CSF), for instance, is the main guide to securing critical infrastructure such as power plants.

As of June 2023, NIST was addressing itself to cyber risks in space. That month, they published a NIST Interagency Report (NIST IR 8441), the Cybersecurity Framework Profile for Hybrid Satellite Networks (HSNs). Like all NIST documents, it's long and very detailed, but the essence of its intent can be found early in the text, which reads, "As space becomes more important to our critical infrastructure, the impact of a cyberattack and the corresponding risk increases. The risk to commercial space operations needs to be understood and managed alongside other risks to ensure safe and successful operations."[20]

The framework's focus on HSNs is not an accident. The creators of the framework, who come from leading aerospace and cybersecurity firms, understand that today's satellite networks are the most vulnerable attack surface in government and commercial space sectors. The framework document speaks to this issue, saying, "There is a need to ensure that these systems are secure, and that the integration of components is done in a manner acceptable to the participating organizations."[21]

The framework document offers several schematics of HSNs, which vary in their configurations. Figure 5.1 is one example. Here, a user/operator interacts with a satellite using a virtual payload control center hosted on a cloud service provider, such as Amazon Web Services (AWS). The control center is linked to a global antenna service provider, which, in turn, interacts with a satellite antenna, a satellite "bus," which is a digital point of connection and, finally, to the satellite itself.

Specifically, as the framework describes, the scope of the HSN encompasses a variety of physical and virtual interfaces, including the following:

- Antenna fields
- Virtual machine (VMs) for formatting commands
- Software-defined elements
- User terminals
- Intermediate ground nodes
- Intersatellite cross links, for example, for linking to a payload hosted on other satellites—required for path redundancy, better resolution, improved communication bandwidth, and so forth

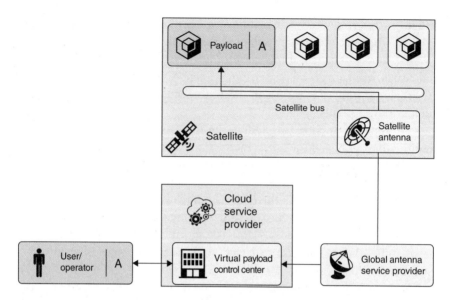

FIGURE 5.1 Example of an HSN with virtualized components

If you're familiar with modern IT, this architecture will make sense. It offers a high degree of flexibility and "loose coupling" that enables efficient, economical satellite operations. There are few hardwired connections. The control center is hosted in the cloud, which gets the satellite operator out of the business of running expensive data centers, and so forth.

While this structure is good for the satellite operator's business, it is vulnerable to cyberattack in multiple places. Each element of the architecture is exposed to cyber threats, as are all the various network connections and software integrations. Each element needs its own countermeasures and cybersecurity practices, such as "patch" updates of software that's revealed to have cyber vulnerabilities. There's a lot of attack surface here.

What's more, in all likelihood, the pieces of this structure will be owned and operated by more than one corporate entity, each with its own cybersecurity policies, tools, and teams. They might even be in different countries. This is not good for security posture. Breakdowns in communication worsen risk exposure.

The HSN cybersecurity framework provides a way to create and implement security controls and countermeasures for this entire ecosystem, and those like it. It does this through a framework "Core" that consists of standards, guidelines, and practices that space five key functional areas of cybersecurity:

1. **Identify** risks to systems, assets, and data.
2. **Protect** digital assets by developing and implementing safeguards—limiting or containing the impact of a potential cybersecurity event.
3. **Detect** the occurrence of cybersecurity events.
4. **Respond** to cybersecurity events and contain their impact.
5. **Recover** from cybersecurity events, maintaining resilience and restoring capabilities or services that were impaired due to a cybersecurity event.

In each of these five core areas, the HSN framework offers numerous recommended policies and controls. For example, under Identify (ID), the framework suggests an Asset Management (AM) set of controls that covers identification of risks affecting assets such as an HSN's physical devices and software platforms. By methodically following the dozens of policies and controls contained in the framework, a cybersecurity team will be able to implement an effective cybersecurity program and achieve strong cybersecurity posture for an HSN.

The HSN CSF represents a solid effort to bring some structure and coherence to the cyber defense of space assets. It's not the only effort in this direction, but its effectiveness remains to be seen, however. Adherence to the framework is voluntary, and a given organization's level of commitment to it is up to them. That is, each entity that participates in an HSN will have to decide how deeply it will implement the framework's controls—and for how long. Lapses in cybersecurity often occur when a once thorough cyber program falls into disuse.

NIST is also developing security standards for the satellite ground segment (NIST IR 8401).[22] This publication outlines how to apply the NIST Cybersecurity Framework to the ground side of COMMSAT operations. Given the risk of attacks on ground stations, this emerging standard seems relevant and promising.

Certainly, the attackers seldom let up. They will seek vulnerabilities and exploit them wherever they can. As a result, space systems remain exposed to cyber risk.

A further encouraging development in space cybersecurity is the creation of ISACs for space. ISACs are common in cybersecurity, enabling stakeholders from different organizations to come together and share threat intelligence and other information that enables them to be collectively better at cyber defense. For example, MITRE's Colorado-based Space ISAC created a Cislunar Affinity Group in late 2023. Their goal is to "raise awareness on the need for solutions in cislunar space in community not traditionally turned towards cislunar, like cybersecurity and the information sharing of vulnerabilities, incidents, and threats."[23]

(Hopefully) Learning from Past Mistakes

The pervasive vulnerability of computer systems and networks is spreading into the space domain. It is likely that cyber incidents in space will grow more serious and costly as the value of space cargoes increases over time. Criminal groups that use cyberattacks to steal and disrupt digital assets on Earth will almost certainly start plying their trade in space, becoming space pirates in the process.

Assuming that this prediction is accurate, we can offer an observation about what has led to this problem, a suggestion on what to do about it, and a warning for future space policy. First, the observation: today's cyber insecurity crisis has many causes, but among the most significant causal factors is a bad combination of insecure engineering and a lack of law enforcement.

The frenetic nature of the tech industry, with its inconsistent (or nonexistent) application of secure design principles, results in digital space assets that embody a range of cyber vulnerabilities. Attackers have remarkable latitude to target space assets across a broad attack surface. Countermeasures and controls can be effective, if they are well deployed, but there is no guarantee of that happening on a consistent basis.

At the same time, there is little to no effective law enforcement that mitigates these threats. While it is possible to prosecute hackers, these cases are meaningless if the hackers are operating from foreign bases. This is a dead certainty when it comes to space cyber piracy. Nations engaged in space commerce will be powerless to intercede as criminal gangs target their space systems.

The "to do" that emerges from this assessment is to develop better security controls for space assets. This is much easier said than done, but it is possible. It seems wise to examine ways to mandate secure design principles for any system that touches on space assets. And, as part of that process, develop and enforce strict policies for what technologies are permitted for use in space.

This is not as far-fetched as it sounds. The way computer systems are certified for use in aviation could provide a model for space. You can't install any old computer to control a 787, for example. Any critical computer system on an aircraft must pass certifications according to standards set by the FAA. It's time to start thinking this way about space.

The warning goes like this: today's cybersecurity disaster offers a painful example of what can happen when critical technologies are allowed to develop by accident. Now is the time to avoid this outcome in space. Decisions that we make in the next few years about computer system designs and cybersecurity controls in space will be with us for a century at least. We will return to this subject in a later chapter.

Takeaways

Space assets, which are to a great extent simply orbiting computer systems, are vulnerable to cyberattacks. Space pirates may use hacking techniques to hijack or disrupt space commerce. In fact, this is already starting to occur, just without ransom demands. This risk is relevant to a discussion of space piracy for several reasons. For one thing, cyber vulnerability should be remediated to better protect space assets from malicious actors. Even if hacking satellites is harder than it looks, as their value grows, so will the temptation to hack them. In the bigger picture, however, the whole current cyber disaster, the remarkable exposure of computer systems, and networks to attack despite hundreds of billions spent very year to prevent such incidents should signal a warning to anyone planning for space security. Today's cyber risks are the results of well-intentioned but terrible engineering decisions made a generation ago. It would be folly to miss the opportunity to learn from these mistakes in space. Organizations like NIST, along with the private sector, are starting to tackle these security challenges for space. The outlook for the enforcement of such standards is still very much up in the air, however.

Notes

1. Brett Tingley, "Spies and Hackers are Targeting the US Space Industry: Report," Space, August 22, 2023, https://www.space.com/us-government-warning-space-industry-satellites-hacking.

2. Saheed Oladimeji and Sean M. Kerner, "SolarWinds Hack Explained: Everything You Need to Know," Tech Target, November 3, 2023, https://www.techtarget.com/whatis/feature/SolarWinds-hack-explained-Everything-you-need-to-know#:~:text=The%20third%2Dparty%20software%2C%20in,detection%2C%20even%20by%20antivirus%20software.

3. Andy Greenberg, "How a Shady Chinese Firm's Encryption Chips Got inside the US Navy, NATO, and NASA," Wired, June 15, 2023, https://www.wired.com/story/hualan-encryption-chips-entity-list-china.

4. Robert C. O'Brien, "NDAA Needs to Tighten National Security Restrictions on Chips Made in China," *Washington Examiner*, January 23, 2023, https://www.washingtonexaminer.com/news/business/2720524/ndaa-needs-to-tighten-national-security-restrictions-on-chips-made-in-china.

5. Craig Sparling and Sandeep Rath, "Akamai Prevents the Largest DDoS Attack on a U.S. Financial Company," Akamai, September 8, 2023, https://www.akamai.com/blog/security/akamai-prevents-the-largest-ddos-attack-on-a-us-financial-company.

6. "Cyber Security Market Size, Share & Trends Analysis Report By Component, By Security Type, By Solution, By Services, By Deployment, By Organization Size, By Verticals, By Region and Segment Forecasts, 2024–2030," Grand View Research, accessed August 9, 2024, https://www.grandviewresearch.com/industry-analysis/cyber-security-market#:~:text=Report%20Overview,12.3%25%20from%202023%20to%202030.

7. Ellen Nakashima and Paul Sonne, "China Hacked a Navy Contractor and Secured a Trove of Highly Sensitive Data on Submarine Warfare," Washington Post, June 8, 2018, https://www.washingtonpost.com/world/national-security/china-hacked-a-navy-contractor-and-secured-a-trove-of-highly-sensitive-data-on-submarine-warfare/2018/06/08/6cc396fa-68e6-11e8-bea7-c8eb28bc52b1_story.html.

8. Rahul Kumar, "Is China Trying to Hack into Satellite Communication as Part of Asymmetric Warfare?," *Indian Narrative*, March 26, 2023, https://www

.indianarrative.com/world-news/is-china-trying-to-hack-into-satellite-communication-as-part-of-asymmetric-warfare-124351.html.

9. Dov Greenbaum, "Who is Going to Stop Space Terrorists?," Calcalis Tech, January 8, 2022, https://www.calcalistech.com/ctech/articles/0,7340,L-3926737,00.html.

10. Martin C. Libicki, *Cyberspace in Peace and War, Second Edition*, U.S. Naval Institute, 2021, 512.

11. Craig Timberg, "Net of Insecurity: A Flaw in the Design," Washington Post, May 30, 2015, https://www.washingtonpost.com/sf/business/2015/05/30/net-of-insecurity-part-1.

12. Timberg, "Net of Insecurity."

13. John Breeden II, "US Issues Threat Warning after Hackers Break into a Satellite," Defense One, August 23, 2023, https://www.defenseone.com/threats/2023/08/national-intelligence-office-issues-cyber-warning-government-and-commercial-satellites/389671.

14. Pierluigi Paganini, "Hacking Satellites . . . Look Up to the Sky," Infosec, September 18, 2013, https://www.infosecinstitute.com/resources/scada-ics-security/hacking-satellite-look-up-to-the-sky.

15. Harrison Caudill, "Big Risks in Small Satellites: The Need for Secure Infrastructure as a Service," Aerospace Research Central, November 2, 2020, https://arc.aiaa.org/doi/10.2514/6.2020-4017.

16. "Space Attacks Open Database Project," Space & Cyber Security, accessed August 10, 2024, https://www.spacesecurity.info/en/space-attacks-open-database.

17. James Middleton, "Tamil Tigers hack satellite," Telecoms, April 13, 2007, https://www.telecoms.com/satellite/tamil-tigers-hack-satellite#close-modal.

18. Maggie Miller, "The U.S. is Getting Hacked. So the Pentagon is Overhauling its Approach to Cyber," *Politico*, December 9, 2023, https://www.politico.com/news/2023/09/12/pentagon-cyber-command-private-companies-00115206#:~:text=%E2%80%9CDOD's%20cyber%20strategy%20was%20extremely,Command%20from%202020%20to%202022.

19. Larisa Redins, "SpiderOak, Space Development Agency Collaborate to Enhance Space Cybersecurity," Edgeir, October 23, 2023, https://www.edgeir.com/spideroak-space-development-agency-collaborate-to-enhance-space-cybersecurity-20231023.

20. "Cybersecurity for the Space Domain," National Cyber Security Center of Excellence, accessed August 10, 2024, https://www.nccoe.nist.gov/cybersecurity-space-domain.

21. Hybrid Satellite Networks (HSN) Cybersecurity," National Cyber Security Center of Excellence, accessed August 10, 2024, https://www.nccoe.nist.gov/projects/hybrid-satellite-networks-cybersecurity.

22. Suzanne Lightman, Theresa Suloway, and Joseph Brule, "Satellite Ground Segment: Applying the Cybersecurity Framework to Satellite Command and Control," Computer Security Resource Center, December 2022, https://csrc.nist.gov/pubs/ir/8401/final.

23. Colorado Springs, "Space ISAC Announces Stand up of Cislunar Affinity Group," Space ISAC, February 28, 2023, https://spaceisac.org/space-isac-announces-stand-up-of-cislunar-affinity-group.

6 The Space Law, Policy, and Treaty Landscape

The signing of the Outer Space Treaty at the United Nations on January 27, 1967. From left to right: USSR Ambassador to the UN Anatoly F. Dobrynin, UK Ambassador Sir Harold Beeley KCMG CBE, U.S. Ambassador Dean Rusk, and U.S. President Lyndon Baines Johnson.

Source: Bettmann/Getty Images

A n old joke asks, "What happens when you cross a lawyer with a mob guy?" The answer is "someone who makes you an offer you can't understand." So it seems to be when it comes to the legalities affecting space crime.

While the inaccessibility and vastness of space make it seem like an environment that's beyond the reach of Earth's law, there are in fact many laws covering human, national, and corporate conduct in space. The problem is that it's not clear which laws will apply—or can be enforced—as the scope of commercial activity in space grows beyond today's predominant sphere of low Earth orbit (LEO).

Will existing legal frameworks for space be adequate to handle crime and piracy in space? This chapter attempts to sort out the various treaties and legal precedents that have been established for space so far, and it examines the gaps that remain. In particular, we will look at laws covering maritime trade and crime, which deal with many comparable scenarios involving the transport of valuable cargoes.

Further to this point, we will explore the complex but well-established interactions between law, finance, and insurance that have characterized maritime trade for centuries—and try to draw parallels with space. We do this because it is likely that existing insurance and financial practices for oceangoing commerce will serve as a readymade legal template for space commerce. Indeed, some of the same players may simply extend their businesses from seagoing trade to space. This is not necessarily a bad idea, but sea law has its limits when applied to space.

Laws and Treaties Relevant to Space Crime and Piracy

The development of laws governing conduct in outer space is akin to the story of a small town that realized it needed a traffic light only after two of its residents bought cars. Before that, no one saw a need for such a contraption. Space, like that mythical one-horse town, had no traffic to speak of until 1958, when the Soviet Union shocked the world with the launch of Sputnik, the first artificial satellite to orbit Earth.

Suddenly, people in positions of authority the world over began to see a need for laws dealing with safe conduct in space as well as fairly sharing space resources. The results comprise a collection of laws and treaties, many of which were formed and continue to be governed by the United Nations (UN).

The Communications Satellite Act and INTELSAT

Four years after Sputnik, in 1962, the U.S. Congress passed the Communications Satellite Act, which established a private entity known as the Communications Satellite Corporation (COMSAT). COMSAT was authorized to develop a global commercial satellite communications system, which it proceeded to do. It oversaw the launch of Intelsat 1, which carried telecommunication transmissions between North America and Europe. COMSAT also facilitated the creation of related satellite infrastructure on the ground to support Intelsat 1 and many subsequent craft.

As use of satellites grew, stakeholders, including national governments and commercial enterprises, saw the need for more international agreements regarding the operation of satellites. In 1964, 19 countries, including the United States, United Kingdom, Canada, and France, formed the International Telecommunications Satellite Organization (ITSO), which deployed a single global commercial communications satellite system. Changing its name from ITSO to INTELSAT in 1971, the members also signed the permanent Agreement Relating to the ITSO.

INTELSAT's guiding principles came from UN Resolution 1721 (XVI). This resolution held that satellite communications should be available to all nations on a global and nondiscriminatory basis. As the agreement stated, INTELSAT's purpose was "to continue and carry forward on a definitive basis the design, development, construction, establishment, operation and maintenance of the space segment of the global commercial telecommunications system."[1]

In practical terms, these agreements give authority to the International Telecommunication Union (ITU), a UN agency that oversees information and communication technologies. The ITU, in turn, is run according to its own Constitution, Convention, and the Radio Regulations. These are the international rules governing satellite telecommunications. ITU rules see to it that satellites are placed into specified orbits and use permitted radio frequency spectrum. The goal is to ensure fair sharing of these limited resources.

Further to this point, each country operating under the ITU rules can choose its own radio spectrum and orbital positions on a "first come, first served" basis. Member states are obligated to avoid harmful interference with one another's radio spectrum for space operations. States are therefore obligated to notify the ITU of their frequency choices, and so forth.[2]

So far, so good, but what if a conflict emerges? Article 56 of the ITU Constitution provides a mechanism for dispute resolution. Arbitration is a path to settling a conflict, among others. The ITU can facilitate bilateral negotiations between states in conflict over orbits and radio spectrum.

However, the ITU does not have any power of enforcement. Compliance is voluntary. There have been disputes over the years, but they have been worked out according to these rules, more or less.

What's not clear is how this genteel "honor system" will work in the event of brazen lawbreaking in space. Chances are, the ITU will not be able to control malicious actors who interfere with satellite communications and orbits for criminal purposes. And when that happens, we'll be in new territory, legally speaking. Remedies do exist, but they have not been tested.

In the meantime, international corporate litigation appears to be the venue where some of these issues will play out. The messy dispute involving Kleo Connect offers a glimpse at the kind of high-stakes space conflicts that are playing out in the legal system. Kleo Connect was a joint venture between German and Chinese investors to create a fleet of LEO communications satellites. The business used satellite licenses from Liechtenstein, a tiny country with big ambitions in space. The backers of Kleo Connect are alleged to have connections to their respective national governments.

The shareholders are now in a bitter lawsuit concerning control over the entity and its satellites. There are charges that the venture was simply a Trojan horse allowing each partner's respective nation-states to influence or control the satellites. American investors with political connections have entered the fray. The matter will play out in courtrooms in Washington and elsewhere because the LEO satellites cross multiple international jurisdictions as they circle Earth.[3]

The Kleo Connect case shows how complicated it can be to deal with legal disputes in an international space commerce business. And, while the facts are not entirely in evidence, it seems that Kleo Connect started out as a legitimate business venture. One can only imagine how difficult it would be to resolve the conflict if one or more of the founding partners was a front for a criminal organization.

The Outer Space Treaty (1967)

By the mid-1960s, it was becoming apparent that space exploration would require some form of international law to govern the activities of spacefaring nations. At that point, the spacefaring club had only two members, the United States and the USSR, but they were the most powerful nations on Earth at the time, so some rules were in order.

So it was that in 1967, the UN passed Resolution 2222 (XXI), which called for a *Treaty on Principles Governing the Activities of States in the Exploration and Use of Outer Space, including the Moon and Other Celestial Bodies*, also known as the "Outer Space Treaty."

Despite its age and generally broad parameters, the treaty remains the primary legal document covering conduct in space to this day.

The language of the treaty reflects the aspirational, visionary sense that people had of space in the 1960s. It opens by saying, "The States Parties to this Treaty, inspired by the great prospects opening up before mankind as a result of man's entry into outer space, recognizing the common interest of all mankind in the progress of the exploration and use of outer space for peaceful purposes, believing that the exploration and use of outer space should be carried on for the benefit of all peoples irrespective of the degree of their economic or scientific development..."[4]

The treaty then goes on to outline a number of critical rules for space. These include the following:

- Calling for freedom to explore outer space, including the moon, for "the benefit and in the interests of all countries, irrespective of their degree of economic or scientific development, and shall be the province of all mankind" (Article I)
- Establishing that outer space "is not subject to national appropriation by claim of sovereignty, by means of use or occupation..." (Article II)
- Requiring that parties to the treaty "shall carry on activities in the exploration and use of outer space, including the moon and other celestial bodies, in accordance with international law, including the

Charter of the UN, in the interest of maintaining international peace and security and promoting international cooperation and understanding (Article III)

- Banning nuclear weapons or other weapons of mass destruction from space (Article VI)
- Creating a duty to rescue astronauts (Article V)
- Creating liability for states who cause accidents related to space accidents, for example, a failed rocket launch (Article VII)

These are the highlights. For our purposes, the main ideas embodied in the treaty are that space exploration should accrue to the benefit of all humanity, that national sovereignty does not exist in space, and otherwise that international law shall apply to space endeavors.

The cooperative spirit of the treaty continues to inspire us more than 50 years later. Space remains a frontier that can bring out the best in humanity. However, the agreement is not suited to the kind of sprawling and rapidly growing global commerce now moving into outer space. Indeed, the words "money," "private," "commerce," "business," and "corporation" do not appear anywhere in the treaty.

It's not that the framers of the treaty could not conceive of private space commerce. They probably just didn't imagine it being relevant anytime soon. Subsequent agreements have made a start at addressing private corporate interests in space, but they are still incomplete.

China has taken initiative on this issue, for one. In 2024, the Chinese delegation to the UN's Working Group on Legal Aspects of Space Resources Activities, submitted a position document that seeks to make the use of space resources legal, while calling for adherence to existing frameworks of international space law, such as the Outer Space Treaty.[5] The Working Group sits under the Legal Subcommittee of the UN's Committee on the Peaceful Uses of Outer Space (COPUOS).

Convention on Registration of Objectives Launched into Outer Space

As satellite launches from multiple nations became commonplace in the early 1970s, the UN decided to enact an agreement that dealt with registering spacecraft for purposes of tracking and accountability. In 1974, the

United Nations Office for Outer Space Affairs (UNOOSA), an office of the U.N. Secretariat tasked with facilitating peaceful international cooperation in space, oversaw the adoption of the Convention on Registration of Objects Launched into Outer Space, which is generally referred to as the "Registration Convention."

Now ratified by 72 nations, the Registration Convention requires states to provide the UN with details about the orbits of any "space objects" it launches:

- Name of launching state
- An appropriate designator of the space object or its registration number
- Date and territory or location of launch
- Basic orbital parameters (Nodal period, Inclination, Apogee, and Perigee)
- General function of the space object

Internally, the United States has its own space vehicle registration program, more or less. The FAA licenses space launches. The FCC licenses the use of radio spectrum frequencies in space, and the National Oceanic and Atmospheric Agency (NOAA) issues licenses for private space-based sensing activities.

Other Notable Agreements

Over the years, other international agreements and policies have come into existence that touch on different aspects of space operations and the space economy. For example, the 1979 Agreement Governing the Activities of States on the Moon and Other Celestial Bodies (the Moon Agreement) "reaffirms and elaborates on many of the provisions of the Outer Space Treaty as applied to the Moon and other celestial bodies, providing that those bodies should be used exclusively for peaceful purposes, that their environments should not be disrupted, that the United Nations should be informed of the location and purpose of any station established on those bodies."[6]

Additional agreements that are relevant to a discussion of space law include the Agreement on the Rescue of Astronauts, the Return of Astronauts and the Return of Objects Launched into Outer Space, the

Convention on International Liability for Damage Caused by Space Objects, the Principles Governing the Use by States of Artificial Earth Satellites for International Direct Television Broadcasting, the Principles Relating to Remote Sensing of the Earth from Outer Space, and Principles Relevant to the Use of Nuclear Power Sources in Outer Space.

One agreement that may become relevant to space piracy is the Convention for the Suppression of Unlawful Acts Against the Safety of Maritime Navigation. Ratified in 1988, the convention is meant to "ensure that appropriate action is taken against persons committing unlawful acts against ships. These include the seizure of ships by force; acts of violence against persons on board ships; and the placing of devices on board a ship which are likely to destroy or damage it."[7]

According to Professor Jonathan Gutoff of the Roger Williams University School of Law, a leading expert on piracy and maritime law issues, the Convention fits the scenario of space piracy. "If you have people in space on one spacecraft, attacking another spacecraft, that's arguably a violation of the convention," he said. "Of course, the specifics of enforcement will have to be seen, but it fits, in legal terms."

Finally, we have the Artemis Accords, the most recent agreement deserving of attention in the context of space law. Drafted in 2020 by NASA and signed by 29 countries, the Accords represent a nonbinding agreement to pursue peaceful exploration of the moon. The signatories are mostly countries that are participating in the Artemis Program, a moon exploration program involving NASA and space agencies from Europe, Israel, and Japan. The Accords adhere mostly to the concepts agreed upon in the Outer Space Treaty. The core tenets of the Accords affirm the primacy of international law on the moon, as well as the goal of peaceful economic development of the lunar surface.

These agreements are mostly national in character. They work at the nation-state level, with state space organizations and private corporations in each signatory nation facing some degree of obligation to observe the various rules set out by the agreements. Issues of jurisdiction and enforcement are subject to international legal mechanisms, which tend to be subjective and weak. As far as deterring major crimes in space, they remain untested.

The Question of Sovereignty and Jurisdiction in Space

The Outer Space Treaty says that space "is not subject to national appropriation by claim of sovereignty, by means of use or occupation..." The matter is not so simple, though. Yes, the United States, as a signatory to the treaty, cannot claim that the moon is now part of the United States. If the United States builds a lunar colony, however, the issue of sovereignty starts to look different.

For clarity on this, and other issues, we spoke with Dr. Frans von der Dunk, a professor of space law at the University of Nebraska. As the recipient of the Distinguished Service Award of the International Institute of Space Law (IISL) of the International Astronautical Federation (IAF) in Vancouver, among numerous space law accolades, Professor von der Dunk has distinctive expertise on this esoteric subject.

"Outer space is an area outside of individual territory sovereignty, so no state can act as if outer space is part of its national territory," explained von der Dunk. "This means that the normal use of territorial jurisdiction and territorial competencies which a state wants to use to fight crime are not going to work in outer space."

He compared the sovereignty and jurisdiction issues in space to those on Earth, saying, "I'm currently a Dutchman in the United States. The American police are entitled to try and prevent any crimes here. If there is suspicion that someone has committed a crime, whether it concerns a U.S. citizen or a Dutchman on U.S. territory, the U.S. authorities have the full right to investigate the case to take me to justice if necessary. That doesn't apply in outer space."

He clarified his position, though, by adding, "That doesn't mean that there is no law in outer space, but for the first few decades of the space age, space has been exclusively the domain of just a few states." While multiple nations now operate in space, von der Dunk explained that "the whole regime of space law is still very much geared to this old situation of states, primarily through the 1967 Outer Space Treaty."

According to von der Dunk, space law views criminal law as a matter for individual states to resolve. "States individually determine what they consider a crime," he said. "And obviously, countries have different opinions on that. In Saudi Arabia, it's formerly a crime to drink alcohol. And in the United States, it's clearly not, just to give you one example." He went on to say, "So that is all seen as falling under individual states."

The paradox of sovereignty in space comes from a detail in the Outer Space Treaty. The USSR, which was effectively the only counterparty to the treaty with the United States, was a communist country in 1967. As such, they didn't want the treaty to provide any basis for commercial endeavors in space. As Professor von der Dunk explained, however, the treaty contains a compromise in Articles VI and VII. As Article IV says, "States Parties to the Treaty shall bear international responsibility for national activities in outer space, including the moon and other celestial bodies, whether such activities are carried on by governmental agencies or by non-governmental entities...."

A nongovernmental entity could be a private corporation. Per the treaty, the nation in which that entity is based thus carries the liability for any problems that entity creates. As a result, the current view of space law is that national sovereignty, at least regarding legal jurisdiction, follows the nationality of the entity that operates in space.

Professor von der Dunk offered an example, saying, "Say, for the sake of argument, that Space X somehow violates certain standards of international law applicable to the moon. In that case, it's the United States that is directly responsible because SpaceX is an American company." A SpaceX vehicle or lunar installation becomes something akin to an embassy: a piece of a sovereign state that exists within another sovereign state, or nonsovereign area, such as outer space.

As a result of this chain of sovereignty, a crime in space could theoretically be prosecuted under the jurisdiction of the entity's home nation. Professor von der Dunk further noted that this framework is more than theory when it comes to the International Space Station (ISS), which was formed through a partnership of 15 countries.

With the ISS, von der Dunk explained, these nations "have basically agreed that if there is something which could be potentially subject to criminal jurisdiction, that states that are involved can actually exercise jurisdiction over such purported crimes." For instance, if an American astronaut on the ISS physically assaults a Russian astronaut, that would be a clear violation of both American and Russian laws. Back on Earth, the American assailant could be tried in an American or Russian court for the crime. The location of the criminal trial would depend on where the astronaut returned to Earth. If he lands in Russia, he could be tried in Russia.

For space law attorney George Anthony Long, the issues of jurisdiction and enforcement in space are rooted in the Outer Space Treaty of 1967. Pursuant to the treaty, crime in outer space can be sourced in international law and state law.

Outer Space Treaty Article III extends international law to outer space. This incorporates conduct deemed criminal under customary international law. Any individual committing one of the customary international law crimes is subject to universal jurisdiction which allows any state to prosecute the person for the criminal conduct. Consistent with the S.S. Lotus principle, conduct of a person that does not constitute a crime under customary international law is not illegal under international law.

As Long explained, the Outer Space Treaty Articles VI and VII allow the enactment of state laws that criminalize certain conduct or acts in or relating to outer space. Article VI delegates to states the responsibility and duty to police the space activities of their nationals' and of nongovernmental actors, while Article VIII recognizes that a spacecraft and those onboard are subject to the jurisdiction and laws of the registry state.

Such state laws, however, will generate issues relating to the reach of jurisdiction and enforcement. These issues leapt from the zone of theory into reality in 2019 with what appears to be the first allegation of a crime committed in space. Anne McClain, an astronaut on the ISS, was accused by Summer Worden, her estranged spouse, of illegally accessing her bank account from a NASA space Internet connection.[8] Worden, a former Air Force intelligence officer, reported the incident to NASA's Inspector General, who opened an investigation. Following the ISS's jurisdictional policy, McClain submitted to a deposition upon her return to Earth. The matter has been resolved with no criminal charges filed, but it shows how space law can actually work in practice.

The issue can quickly get messy, however, when the question of responsibility crosses from government to corporate ownership of a space asset. As Professor Margaret Sankey of the Air University put it, "Are the sovereign countries where space businesses are based responsible for the behavior of those companies' assets? What do you do if a commercial satellite is hijacked and is used to dazzle another country's military satellites? How would you go about apportioning blame and trying to figure out how to shut it down, especially if you're dealing with a company that that operates transnationally?"

We will return to these issues in Chapter 9, which deals with commercial impacts of space piracy. In that context, potential disputes over property rights in space become a serious potential problem.

The Need/Temptation to Draw on Maritime Laws and Traditions

The ISS example shows that there is a path to enforcing laws against piracy in space, at least in theory. In practice, no one knows how such a case would actually proceed. A body of law exists, however, that offers some direction on how space law and criminal law enforcement could work in space. That is the law of the sea.

The world's oceans constitute a comparable legal environment to space. As Austin Albin, a former Army intelligence officer, Space Force contractor, and consultant, explained in an article in *The Space Review*, "The UN Convention on the Law of the Sea (UNCLOS) could serve as a model for space-specific international law. There are several similarities between international waters and space, such as freedom of navigation, safety precautions for vessels, and resource extraction."[9]

It's not a particularly auspicious picture, but an examination of maritime law and traditions can be useful in thinking through how law enforcement and antipiracy measures might work in space. Many of the same issues and challenges exist on the high seas.

For insights into maritime law and its parallels with space, we spoke with Professor Salvatore Mercogliano, Chair of the Department of History Criminal Justice and Political Science at Campbell University. Professor Mercogliano is considered one of the world's foremost experts on maritime law and history.

As he explained, once a vessel goes past the 12-mile national limit, and from there, past the 200-mile exclusive Economic Zones, it's in what Professor Mercogliano referred to "the lawless sea." As he said, "It's the outlaw sea. I mean, there's very few laws out there except for the flag of registry of your ship. So, in this way, you have a compelling parallel with space, though I actually think space is maybe in a little better position."

In his maritime security class, Professor Mercogliano runs a scenario past his students: "What stops me from registering a vessel in a country, going out and fishing up a marine cable, an underwater Internet cable, getting on a Facebook call with the owner of the cable and saying, 'I'm going to cut this in half, unless you give me several million dollars?' What law am I breaking?" (This threat went from theoretical to real in 2024, when telecom firms operating in the Middle East expressed concern that a Telegram channel supporting the Houthi rebels in Yemen, which had been attacking shipping in the Red Sea, were also threatening to destroy cables running under that waterway.[10])

He added, "Really, the only law enforcement agencies that are relevant are those that cover the ship, in theory—from wherever I happen to be registered, if I bother to register it anywhere. It's international waters. You do whatever you want in international waters."

Professor Mercogliano's provocative example aside, there are, in fact, some laws on the high seas. Maritime law relies on several treaties, the most important of which is the 1982 UNCLOS convention, which was an expansion of the 1958 Convention on the High Seas. It established the right of innocent passage in territorial waters.

Regarding crime at sea, UNCLOS says, "The criminal jurisdiction of the coastal State should not be exercised on board a foreign ship passing through the territorial sea to arrest any person or to conduct any investigation in connection with any crime committed on board the ship during its passage, save only in the following cases: (a) if the consequences of the crime extend to the coastal State; (b) if the crime is of a kind to disturb the peace of the country or the good order of the territorial sea; (c) if the assistance of the local authorities has been requested by the master of the ship or by a diplomatic agent or consular officer of the flag State; or (d) if such measures are necessary for the suppression of illicit traffic in narcotic drugs or psychotropic substances."[11]

UNCLOS Article 101 defines piracy as "any illegal acts of violence or detention, or any act of depredation, committed for private ends by the crew or the passengers of a private ship or a private aircraft...on the high seas against another ship or aircraft."

A coastal state can intervene in a criminal matter at sea under certain circumstances. However, outside the 200-mile limit, this right does not apply. It is theoretically possible for, say, an American sea captain to enforce American law on a ship with American registry. The captain might even summon an American law enforcement agency to the ship to make an arrest. This scenario is highly unlikely, however, partly because most ships are not registered in jurisdictions with strong law enforcement capabilities—and that may be a factor in their favor.

It is possible to see how this legal framework could work in certain space scenarios. For example, if a Canadian spacecraft carrying illegal drugs docks at an American lunar base, the Americans on that lunar base could in theory invoke some space variant of UNCLOS and arrest the Canadians, who are in their "coastal area," so to speak.

Professor Mercogliano thinks space law might be more effective than maritime law partly because space is a newer domain. "Maritime law goes back 1,000 years," he said. "It's ridiculous. You're literally dealing with precedents that date back to Romans."

In his view, maritime law is convoluted due to the uncertainty created by old precedents, national laws, and the lawless sea. Further complicating the situation is the hierarchy put in place by the International Maritime Organization (IMO), the UN's shipping arm. The IMO oversees UNCLOS and creates international conventions that set rules for pollution, safety at sea, and so forth.

"All the signatory nations are represented in the IMO," Mercogliano said. "And sometimes those nation-states' representation is kind of weird." Because, for example, Panama has the same representation as China in the IMO. But they're not equals. They're not equals in many ways.

In his opinion, with space, "There is that potential to legislate it and create a coherent set of rules." Even while saying that, however, he then remarked, "The problem in space is that you've got new players coming in new nations with new laws. And again, whose rules and laws are we going to follow?"

Peter Cook, a former Royal Marine who was active in antipiracy programs at the Horn of Africa and now works as a maritime security consultant, also sees a parallel between space law and maritime law. He said, "If you take the oceans of the world as your starting point for formulating legislation and regulations for space, I think that will be the most suitable model. Inevitably there will be differences; you work your way through it, but I think it's a better model than land-based laws."

As he shared, "Everybody can see borders on land, but you can't see borders at sea; you can't see borders in space. It's comparable terrain, so to speak." Cook then elaborated on the stakeholders in maritime law and commerce and how they are similar to what's expected in space.

At sea, as Cook described, there are three main stakeholder groups that are essential to the commercial shipping industry: the ship owners, the ship manager and the charter, and marine insurance companies. The industry refers to the ship as "the box," because a cargo ship is just that, a box that transports goods from one place to another. In his view, space commerce will evolve along a similar path.

There will be "boxes" that ferry people and goods back and forth into space. They will be owned by investors and corporations and operated by other companies under contract. They may be registered in nations that offer space-based "flags of convenience." Insurance companies, most likely candidates being those that are already experts in maritime insurance, will issue policies to cover losses.

It's quite possible that the same players who now dominate maritime trade will come to dominate space trade. The dynamics and corporate relationships are analogous. So, it makes sense that the legal frameworks governing space commerce and crime will be comparable to those now in effect at sea.

Maritime Insurance Practices and Their Applicability to Space

An appraisal of laws that might deal with space piracy should take into account current maritime insurance practices. While it's impossible to predict what's going to happen with insurance in space, it seems likely that existing modes of maritime insurance will set the pattern for space commerce. This matters because insurance is a critical enabling factor for maritime trade. Nothing moves on the world's oceans without insurance, and space will almost certainly be the same.

Peter Cook said that, in maritime trade, "Insurance and the law are intertwined like strands of a rope. Imagine, he said, a ship is going to transport a bulk ore cargo from Australia to China. The ship owner, which is based in London like so many others, approaches a broker and gives him all the details of the ship and the planned voyage."

The broker discusses the voyage with insurance underwriters, perhaps starting with Lloyds of London. Lloyds might insure 20% of the risk. That would make Lloyds the "lead underwriter." Other underwriters will then assume their own respective portions of risk, maybe 10% or 20% at a time—until the entire cargo is insured.

The maritime insurance business, which is mostly based in the United Kingdom, has specialists for different kinds of vessels and cargos. Some focus on tankers. Others insure bulk carriers or container ships.

However, it's important to note that insuring the cargo is not the same as insuring the ship itself. The ship owner needs a separate policy for the ship and its mechanical systems. These policies may not cover piracy or acts of war. That's a separate policy, too, or a rider on the cargo or boat policy.

Furthermore, the process of negotiation between the pirates and the insurance company is indirect. According to Cook, Lloyds and companies like it do not want to be seen as willing to pay off pirates on demand. Rather, the ship owner must negotiate with the pirates, and then if they want to recoup the ransom they paid, they must file a claim with Lloyds—who will then assess whether or not they will pay it.

Details matter in this context. If the pirates took the crew hostage, that might be covered under an insurance policy, or it might not. The policy might only cover piracy affecting the cargo, not the crew, and so forth.

According to Cook, this insurance ecosystem will most likely adapt itself to space commerce as that field evolves into a major business. "Space is just another unusual area of risk, which they've dealt with before in the last few centuries."

The Issue of Enforceability

If space law feels a bit subjective and hard to pin down, thinking about space law enforcement will be even more bewildering. As Professor von der Dunk explained, a violation of space law is not like a legal matter inside a nation-state. In the United States, for instance, a prosecutor can bring charges against an accused criminal in court. This does not work in space, so far.

Is space an inherently lawless environment? It may turn out that way. For example, Austin Albin said, "The failure of the international community to answer key regulatory questions before space resource exploitation begins in earnest will result in an industry prone to tension and lawlessness....The lack of international law for the emerging space economy will result in disputes, and the growing militarization of space threatens to create an environment of lawlessness and possible armed conflict."[12]

Legal matters in space are resolved in the international community, Professor von der Dunk said. "We do have a World Court, which could potentially address those space legal disputes. However, the court can only decide on disputes where both states have agreed to the jurisdiction of the court to do so. And, of course, the bad guys will not have done that."

Extradition won't work, either, in his view. He cited a court case in the Netherlands against Russia for its alleged involvement in the shooting down of a Malaysian airliner in 2014. Two thirds of the passengers on board were Dutch. "Obviously, the Russians are not going to hand over the people who are charged under Dutch law for committing murder of about 200 Dutch citizens and others. I can foresee similar dysfunctions in space criminal prosecution."

From his perspective, enforcement of space law is subject to political forces outside of the law. "It's probable that issues like space piracy will come down to who is stronger. Like, as we have seen in the world, the stronger you are, the more as a state you can get away with. Iraq couldn't get away with invading a sovereign country like Kuwait 30 years ago. In contrast, Russia so far can get away with invading Ukraine because it is such a powerful country. That's the reality of the international community, a reality that will almost certainly apply to space."

Professor Margaret Sankey of the U.S. Air University had a similar insight, pointing out the dilemma over kinetic military response as an obstacle to enforcing laws related to space crime. In some cases, she said, such as when the Tamil Tigers hijacked a communications satellite, a diplomatic solution was effective. "We didn't send troops to Sri Lanka," she noted. "But, the issue got resolved."

Other times, kinetic response is off limits for a variety of reasons. Cyberattacks, in her view, present an analogous, difficult "gray area." For instance, Russia routinely hacks critical infrastructure in the Baltic states, such as Estonia. "Should this kind of incident trigger NATO Article 5," Sankey asked. "This would precipitate a military response by NATO members to protect their member states. No one thinks so, and that's probably for the best, but the boundary line of action versus inaction continues to creep toward increasingly uncomfortable decisions. The truth is that our agreements aren't built to respond to this properly."

Episodes of space crime will likely present a comparable dilemma over the use of force. According to Sankey, criminal groups might set up launch and recovery facilities in remote, inadequately governed equatorial zones such as in east Africa. There, we might actually see a kinetic military response to space piracy, especially if citizens of Western nations are taken hostage.

Why? Because, as she explained, the balance of power lowers the risk of bad outcomes for the intervening nations. If the United States sends troops to an east African state where a criminal gang—which may have more actual power and control than the official government—is harboring space pirates, the geopolitical ramifications of such a move would be relatively minimal. Alternatively, if a country like Mexico were providing a safe haven for space pirates, it would be far more difficult for the United States to intervene militarily. The stakes are a great deal higher on the international stage.

Professor Mercogliano had a comparable take on the importance of national interest in calibrating enforcement responses to piracy. Citing the Somali pirates' 2009 seizure of the Maersk Alabama and the U.S. Navy's subsequent intervention to save the ship, and Captain Phillips, a story that got turned into a movie, he said, "You can't let Tom Hanks get kidnapped, so you have to intervene. But what they don't show you in the movie is the hundreds of other mariners who were not rescued because their ships didn't fly an American flag. They flew Panamanian flags and so forth. In those cases, the ship owners had to negotiate with the pirates and pay them off. We will probably see similar national dynamics play out in space."

Professor Mercogliano also believes that nongovernmental entities will play a role in space law enforcement, as they currently do in the maritime setting. He offered the example of Sea Shepherd, a nonprofit marine conservation organization that engages in direct action campaigns like crashing into whaling ships to defend wildlife. "We may see similar entities in space," he said. "In the same way that nongovernmental fishery enforcement groups are being hired by countries that lack a coast guard of their own, the world could see nongovernmental space protection organizations active in mitigating space crime and piracy."

Further to this point, Professor Mercogliano alluded to the use of open-source intelligence today in tracking transfers of oil and gas that's under sanction. TankerTrackers, an example he shared, keep track of oil tanker movements. It can reveal shipments that are occurring in violation of international law. A similar setup could emerge to track space cargoes.

Loretta Napoleoni, a journalist who has covered organized crime for decades, raised the issue of government complicity in space piracy, with all of its attendant risks. "Law enforcement is a government responsibility," she said, "But as we have seen for decades with major criminal organizations, they can only function if they enjoy protection from the government. Hardcore, sophisticated piracy cannot operate without links with political power. If we are going to have large international criminal groups engaging in space crime and piracy, they will almost certainly be doing so with the secret permission of various national governments. That will make law enforcement effectively nonexistent in space."

The Potential for "Flags of Convenience" in Space

If space commerce evolves according to the maritime model, which is probable given the similarities in trade patterns, the world may have to deal with space vehicles flying "flags of convenience" as do ships on the high seas. The term "flag of convenience" refers to the common practice of registering a ship in a country other than where its owners and operators are located. For example, a ship owned by an American corporation might register it in Panama and fly the ensign (flag) of Panama.

Reasons for registering a ship in a different country vary, but they mostly have to do with avoiding certain taxes, maintenance costs, and labor and environmental regulations. The practice, which began in the 1920s, was reformed to some degree in 1982 after the catastrophic 1978 Amoco Cadiz oil tanker, registered in Liberia, ran aground off the coast of France and discharged millions of gallons of crude oil. Various international agreements have established some rules governing flags of convenience, but gaps in safety, tax enforcement, work conditions, and so forth remain.

The legal principle that's most relevant to a discussion of flags of convenience in space is the Convention for Registration of Ships' requirement that a flag state be linked to its ships either through an economic stake in the ownership of its ships or by providing mariners to crew the vessels. This is known as "linkage."

The Convention reads:

Every State shall take such measures for ships under its flag as are necessary to ensure safety at sea with regard, inter alia, to:

(a) The use of signals, the maintenance of communications and the prevention of collisions.

(b) The manning of ships and labour conditions for crews taking into account the applicable international labour instruments.

(c) The construction, equipment and seaworthiness of ships.

Each State is required to conform to generally accepted international regulations, procedures and practices and to take any steps which may be necessary to secure their observance.

Professor von der Dunk has written an authoritative paper on the potential for flags of convenience in space and their relationship to existing maritime regulations and practices.[13] Regarding the concept of linkage described in the Convention, he wrote, "In the absence of political feasibility to derogate by way of an international treaty from the sovereign right of a state to determine registration conditions, by listing some key parameters on the international level and requiring states to abide by them it was hoped that the genuine link would translate into genuine concern for the well-being of the ship, the crew and the cargo, as well as for others possibly harmed by their operations, and hence would translate into the effective exercise of jurisdiction and control by way of serious and high-level requirements being imposed upon them and enforced as appropriate."

Let's assume, for the sake of argument, that there is indeed a "genuine link" between national registry and a "genuine concern for the well-being of the ship, the crew and the cargo, as well as for others possibly harmed by their operations." This is debatable, but the existing rules are better than nothing. They do ensure a degree of safety and environmental protection for ships and the world's oceans. They facilitate law enforcement at sea, to some extent.

Will such a linkage become a rule for space vehicles? As Professor von der Dunk explained in his paper, no such linkage exists in the current frameworks. He wrote, "Whilst space law also knows the concept of registration of space vehicles and even has an international treaty providing for the baseline details in the form of the Registration Convention, that treaty does not provide for much by way of either 'genuine link' requirements or specific requirements addressing potential safety concerns."

Article IV of the Registration Convention reads:

> "1. Each State of registry shall furnish to the Secretary-General of the United Nations, as soon as practicable, the following information concerning each space object carried on its registry:
>
> (a) Name of launching State or States.
>
> (b) An appropriate designator of the space object or its registration number.
>
> (c) Date and territory or location of launch.
>
> (d) Basic orbital parameters, including: (i) Nodal period; (ii) Inclination; (iii) Apogee; (iv) Perigee.
>
> (e) General function of the space object.
>
> 2. Each State of registry may, from time to time, provide the Secretary-General of the United Nations with additional information concerning a space object carried on its registry.
>
> 3. Each State of registry shall notify the Secretary-General of the United Nations, to the greatest extent feasible and as soon as practicable, of space objects concerning which it has previously transmitted information, and which have been but no longer are in Earth orbit."

As he put it, "Within the corpus *juris spatialis internationalis* there is no reference whatsoever to certification of spacecraft, requirements with respect to crew or the general safety of operations, other elements prominently involved in implementing a 'genuine link' requirement through substantial legal and factual control."

In his view, the closest we get to dealing with linkage is a clause in Article IX of the Outer Space Treaty, which reads, "States (...) shall conduct all their activities in outer space, including the Moon and other celestial bodies, with due regard to the corresponding interests of all other States (...) and conduct exploration of them so as to avoid their harmful contamination (...) and, where necessary, shall adopt appropriate measures for this purpose." This is rather vague and, while helpful, is quite removed from anything resembling an effective registration rule that links to responsibilities for safety and respect for any kind of rule of law in space.

Professor von der Dunk is primarily concerned with legal liability for accidents, such as rocket crashes, but his general conclusion is that flags of convenience are likely to emerge in some form in the realm of space. Specifically, the existing treaty frameworks allow for each launch state to define its own space liability laws—opening the door for differences in potential penalties for accidents and comparably damaging scenarios.

If he's right, flags of convenience will be common in space. Professor Margaret Sankey of the Air University concurs. She offered the example of Luxembourg, whose banks are now offering space vehicle registration in a way that's comparable to Panamanian registration for ships. "If you're not happy with Luxembourg as a flag of convenience in space, how, exactly do you punish Luxembourg?" she asked. Good question....

This will have implications for space crime and piracy. Flags of convenience at sea are an enabling factor in smuggling. They also create a divergence of interest between the owner of a vessel and its crew, making it easier for pirates to hold crews hostage; for example, an American ship owner may not feel very motivated to pay a ransom to free a crew from Pakistan. Similar problematic dynamics could arise in space with flags of convenience.

Takeaways

Most people would agree that crime, piracy, and smuggling in space are wrong, at a minimum, and should be against the law. But, what law? Space is not a lawless environment, but the laws of nation-states do not readily apply. A few major agreements, such as the Outer Space Treaty, establish a baseline for international law governing activities in space. However, it is not clear how the treaty and various follow-on agreements and conventions would deal with actual piracy in space. Sovereignty, jurisdiction, and law enforcement are murky, at best. Maritime law offers precedents that may be useful in working out how law enforcement and antipiracy efforts will work in space, but a lot of work remains to be done. Problems like "flags of convenience" threaten to make antipiracy measures complicated to execute.

Notes

1. "Agreement relating to the International Telecommunications Satellite Organization 'INTELSAT' (with annexes)," Opened for signature August 20, 1971, *UN Treaty Series* 1220, no. 19677 Art. 2 (March 27, 1981), https://treaties.un.org/doc/Publication/UNTS/Volume%201220/volume-1220-I-19677-English.pdf.

2. Elina Morozova and Yaroslav Vasyanin, "International Space Law and Satellite Telecommunications," Planetary Science, December 23, 2019, https://doi.org/10.1093/acrefore/9780190647926.013.75.

3. Eleanor Olcott, "The Corporate Feud over Satellites that Pitted the West against China," *Financial Times*, June 22, 2022, https://www.ft.com/content/f1f342ab-d931-44ca-bf0d-f7762db76982.

4. "Treaty on Principles Governing the Activities of States in the Exploration and Use of Outer Space, including the Moon and Other Celestial Bodies," Opened for signature December 19, 1966, *Resolution Adopted by the General Assembly,* no. 6431. https://www.unoosa.org/oosa/en/ourwork/spacelaw/treaties/outerspacetreaty.html.

5. Andrew Jones, "China Outlines Position on Use of Space Resources," *Space News,* March 6, 2024, https://spacenews.com/china-outlines-position-on-use-of-space-resources.

6. UN General Assembly, *Agreement Governing the Activities of States on the Moon and other Celestial Bodies*, 34h sess., Res. 34/68. https://www.unoosa.org/oosa/oosadoc/data/resolutions/1979/general_assembly_34th_session/res_3468.html.

7. "Convention for the Suppression of Unlawful Acts Against the Safety of Maritime Navigation, Protocol for the Suppression of Unlawful Acts Against the Safety of Fixed Platforms Located on the Continental Shelf," International Maritime Organization, Opened for signature, March 10, 1988, *SUA Treaties* (October 14, 2005), https://www.imo.org/en/About/Conventions/Pages/SUA-Treaties.aspx.

8. Mike Baker, "NASA Astronaut Anne McClain Accused by Spouse of Crime in Space," *New York Times*, August 23, 2019, https://www.nytimes.com/2019/08/23/us/astronaut-space-investigation.html.

9. Austin Albin, "The International Community is Not Prepared for a Future in Space," The Space Review, September 5, 2023, https://www.thespacereview.com/article/4642/1.

10. Patrick Wintour, "Houthis May Sabotage Western Internet Cables in Red Sea, Yemen Telecoms Firms Warn," The Guardian, February 5, 2024, https://www.theguardian.com/world/2024/feb/05/houthis-may-sabotage-western-internet-cables-in-red-sea-yemen-telecoms-firms-warn.

11. "United Nations Convention on the Law of the Sea," International Maritime Organization, accessed August 10, 2024, https://www.imo.org/en/OurWork/Legal/Pages/UnitedNationsConventionOnTheLawOfTheSea.aspx#:~:text=The%20United%20Nations%20Convention%20on,the%20oceans%20and%20their%20resources.

12. Albin, "Law of the Sea."

13. Frans G. von der Dunk, *Towards 'Flags of Convenience' in Space?* University of Nebraska, 2012.

7 Criminal Organizations That Might Pursue Space Piracy

Mexican army soldiers arrive in Sinaloa to reinforce security activities derived from the confrontation between armed forces, police, and members of organized crime, October 2019.
Source: parrazurita/Shutterstock

Who will be the space pirates, if anyone? We've discussed this earlier, but now we want to go into more depth. With the ongoing caveat that we may be mistaken and that there will, in fact, be no such thing as space piracy, we wanted to explore the most likely candidates. In this, we are constrained by a paradox of the present versus the future. We feel compelled to keep our analysis rooted in present-day reality even as we forecast future events that may turn out quite differently from what we imagine.

We need to think about the long run, but the further out we get in the future, the more fantastic the scenarios can get. In the year 2573, Coneheads from the planet Remulak could rebel and become space buccaneers, but that will not help us understand the risks we face in the near term. The predictions in this chapter are based on current criminal activities, because, as John Maynard Keynes said, "In the long run we are all dead."

Criminal Cartels, the Leading Candidate for Space Piracy

Who is in a position to engage in piracy in space? The answer will change over time. In the future, as space travel becomes less expensive and complicated, the barriers to entry for space pirates will become lower. As of now, anyone considering space piracy would need to have access to significant capital, technology, and expertise. For us, this means today's large-scale criminal organizations.

Our perspective is borne out by the prevalence of criminal organizations backing pirates in the past and present. In the eighteenth century, seafaring pirates were able to function because they had a support system on land that provided them with safe havens and markets for stolen goods. More recently, Somali pirates functioned within the protective zone of land-based warlords with cartel-like organizations. And, as noted in Chapter 2, Mexican cartels engage in piracy involving the theft of oil cargoes in the Gulf of Mexico.

Before we go further into this discussion, we want to make two comments. One is that we are aware that we may have developed tunnel vision on this issue. We view cartels as uniquely suited for space piracy, but they are not the only organizations that can pursue this type of crime.

Second, while we focus on Mexican cartels in this chapter, this should not be taken as a criticism of Mexico or Mexican people. The alleged Mexican criminal threat has been a hot button in American politics in recent years, and it is not our intention in any way to perpetuate that idea.

Mexico simply offers a good, nearby example of how cartels function. As we will explain shortly, such organizations exist all over the world. The Yakuza of Japan are but one of many examples.

What is a criminal cartel? Definitions vary, but the general consensus is that a cartel is a large criminal organization that dominates a criminal category in a geographic region. Cartels often have influence or outright control over law enforcement agencies that are supposed to police them, as well as connections and influence in the government. They are monopolistic in nature, typically employing extreme brutality and violence to protect their exclusive control over criminal rackets.

How is a cartel different from a gang? The two organizational types are similar, but a cartel is bigger. It manages extensive criminal operations across regional and national borders and launders its money through sophisticated techniques.

Examples of Cartels

Cartels or equivalent criminal organizations operate in nearly every corner of the globe. In Latin America, there are Columbia's Medellin and Cali Cartels, along with Mexico's Jalisco New Generation Cartel, the Sinaloa Cartel, and Los Zetas. In Asia, where cartels may be referred to as "Triads," there are the Golden Triangle and Golden Crescent. Europe is home to many large-scale organized crime enterprises, including the Sicilian Mafia and criminal gangs from Russia, Albania, Serbia, and many others. These are just a few examples. There are actually hundreds of serious, sizable criminal organizations currently active in nearly every country on Earth. Their level of power and influence depends on the relative strengths of governments in the places where they operate.

Additionally, cartels sometimes work together across territories. One example of this scenario arose with Joaquín Archivaldo Guzmán Loera, known as "El Chapo," the former leader of the Sinaloa Cartel, who expanded into the cocaine business in China and Hong Kong through a cooperative relationship with the Triads. A report from the Brookings Institution reflected on this arrangement, saying, "In a distant land of vastly different culture and language, the triads became an indispensable interface to consumers for the Sinaloa Cartel. Moreover, many triads had multifaceted linkages and various forms of protection from the Chinese Communist Party (CCP) and Chinese government officials."[1]

Conversely, as the report further declared, the cooperation can go in the other direction: "The triads sometimes help promote Chinese economic interests abroad and serve as informal enforcers against Chinese diaspora members who threaten the interests of the CCP or Chinese government."

How Cartels Operate

Cartel activities vary greatly by region and organization. In some cases, a cartel's main purpose will be the trafficking of drugs, people, or both. Alternatively, or in addition, many cartels engage in local extortion and other parasitic, extractive activities against the populations of their home countries.

For insight into the world of cartels, we spoke with Nathan P. Jones, PhD, a nonresident scholar in drug policy and Mexico studies at the Baker Institute and an associate professor of security studies at Sam Houston State University in the College of Criminal Justice. Professor Jones is considered an expert on Mexican cartels.

He offered an example of how a cartel might extort money from a local business. It might be a simple threat, as in "Pay us 10% of your revenue or we'll kill your family." Or, and this is more often the case, it's a threat masked by a suggestion that the business buy from some designated vendors, which are controlled by the cartel. The vendor charges highly marked up prices, which enables the cartel to profit handsomely from the relationship. The target business knows that they will be subject to violent retaliation if they don't comply.

For Professor Rodrigo Canales of the Yale School of Management, cartels embody a business model that includes branding and organizational structure. As he explained in a TED Talk, a cartel might arrive in a locality and "let people know that they are there, and they go to the most powerful local gang and they say, 'I offer you to be the local representative of the Zeta brand.' If they agree—and you don't want to know what happens if they don't—they train them and they supervise them on how to run the most efficient criminal operation for that town, in exchange for royalties."

Cartels and Governments

Cartels tend to have close connections to governments where they operate. Sometimes, these connections can invert the standard law enforcement model—to the point where it is the criminals who control what the government does, rather than the government cracking down on criminal activities. Professor Gaudalupe Correa-Cabrera of George Mason University, author of *Los Zetas Inc.: Criminal Corporations, Energy, and Civil War in Mexico*, provided some useful insights in this regard.

In her work, she found that cartels act like transnational corporations. With that mode of operation, cartels form links with state structures and the people who work for the state. While she cautions not to oversimplify what is almost always a complex set of relationships, she shared an example of this phenomenon: "Criminal paramilitary groups are often connected with the state in one way or the other one. So, you have the Mexican government fighting a war on drugs, but they are aligned with some groups like the Jalisco cartel that are in the drug trafficking business."

Or, she said, "The Zacatecas cartel works with the Mexican Marines and Mexican Navy, as well as with the government of Veracruz to get rid of Law Centers." From there, the boundary between military service and criminal activity is porous. As exemplified by Los Zetas, former elite, U.S.-trained soldiers get paid very well to bring their discipline and advanced tactics to the service of a criminal cartel. The results can be deadly in the extreme.

How do cartels achieve this influence over organs of government? Again, simple answers should be avoided, but in general, cartel control over governments arises from a combination of corruption and intimidation. People with authority in the government may be bribed, typically in sums that are far out of reach for them on their government salaries. Or they can be threatened with violence if they don't comply with the cartels' wishes.

Alternatively, as Professor Nathan Jones explained, former government employees find themselves adrift and are seeking a meaningful, lucrative new career. "Let's say you retire, based on a mandatory age limit, from the Space Force or Air Force. Some people adapt well and find an interesting new path in life. Others do not. They may resent being out of work and look favorably at the opportunity to use their skills for a criminal purpose. Believe me, this happens a lot more often than you might imagine."

The results are the same, either way. The cartels become capable of compromising government processes. For example, cartels involved in human and drug trafficking have paid off U.S. Border Patrol officers at the Mexican border. In just one of many such cases, a drug smuggler, presumably in the employ of a large criminal organization, paid a Border Patrol agent to "wave through" cars containing cocaine at the Falfurrias checkpoint in Texas. He was sentenced to 10 years in prison for this crime, but there undoubtedly other candidates to take over what he was doing.[2]

Cartels may not even need government help to operate inside of U.S. government areas of protection. A 2023 article alleged that cartels like the Sinaloa and Jalisco organizations had deployed "Intelligence Bunkers" near the U.S.–Mexico border, complete with spy cameras, to keep an eye on their operations and what U.S. and Mexican authorities are doing to interdict their trafficking. Mexican officials found more than a dozen such cameras near the border in Tecate, Mexico.[3]

For Professor Nathan Jones, the surveillance can go in two directions. He said, "In some ways, it's neo-feudal. It's almost like the feudal warlord lord models of social control in some of these rural areas. But, it's also combined with incredible potential for ubiquitous surveillance and control over local populations. You're finding that cartels have command rooms where they're finding that there are cameras set up throughout the city. And they're watching everyone that's going everything that's going on."

Cartel-Driven Space Piracy Scenarios

Piracy is a form of crime pursued by at least some cartels. For example, as mentioned earlier, Mexican cartels are engaged in piracy against oil rigs and oil cargoes in the Gulf of Mexico. According to Ioan Grillo, a journalist based in Mexico who as reported on oil piracy by cartels, the move to piracy represents a diversification from drug smuggling, which had become too dangerous for many players. As he put it, "As the drug war raged, gangsters ramped up other crimes from stealing crude oil to wildcat mining to good old piracy. The cartels have moved into a portfolio of these rackets."[4]

Will cartels expand their portfolio of rackets, so to speak, to space as the space economy grows? Given that cartels seem to be able to operate pretty anywhere they want and do whatever they want, often with the local government looking the other way or actively helping them, the answer is probably "yes." Not everyone agrees with this prediction, however.

Furthermore, space piracy can include criminal activities on Earth, such as stealing space cargoes once they've landed. In this context, existing criminal organizations might easily adapt to this new venue for plunder.

Can Cartels Actually Operate in Space?

Can today's criminal cartels operate in space? As of now, that's a "no," but things could change pretty quickly. Having a sense of imagination is useful in thinking through what might happen. A few times, when we asked experts about their views on cartels as space pirates, they expressed significant incredulity. "Are you saying that a drug trafficker is going to build a rocket and fly into space to hijack satellites? That seems hard to believe." It is, and that's not what we're talking about.

If you ask the question differently, such as wondering if today's cartels could gain influence over a launch company and threaten to destroy a satellite on the launchpad, then the answer is almost certainly "Yes, cartels can engage in piracy that affects the space industry." They already have ample reach, capabilities, and money to perpetrate such an act of piracy.

Kidnapping/Devicenapping

Kidnapping of people or hardware in space seems like a probable scenario for cartels engaging in space piracy. It would match their existing tactics on land and sea. Cartel scholars like Nathan Jones can envision cartels either acquiring or gaining control of technology that could impair satellites in orbit.

He said, "There's a lot of opportunity there for organized crime to potentially play spoiler and run protection rackets that cause real problems. It may be some years off, but I can imagine a cartel being able to position (or reposition) an object in space that could damage or destroy an expensive satellite—as a way to extort money out of its owner. It's not all that different from what they're doing on Earth right now."

One possible path into such activity was revealed in a seemingly minor story of infiltration of airports by criminal gangs. As reported in the *Wall Street Journal*, airline and airport employees are being recruited to smuggle drugs and traffic in stolen goods in U.S. airports. In some cases, their tasks originate abroad, such as in the case of an airport worker in the U.S. Virgin Islands handing almost five pounds of cocaine to a smuggler who took the drugs on a flight to Florida.[5]

This may not sound like a big deal, but it's important to understand that if criminal gangs, almost certainly affiliated with the major cartels, can infiltrate airports, they can infiltrate space ports. With few current security standards in space commerce, it might even be easier for them. With compromised employees in place at space ports, cartels are in a strong position to engage in kidnapping of space tourists and space workers or the hijacking of space hardware.

For "Bob," the retired intelligence officer, the mitigation of this type of piracy will be challenging for the governments of target space businesses. As he explained, "The cartels are located outside the borders of the United States. As a result, we have limited jurisdiction in enforcing any kind of law on them. So, our capability would be very minimal. We could inveigh upon the government of Mexico to interfere or intervene on our behalf. But, we would have no capability unless the Mexican government offered it to us to go there."

Infiltration or Takeover of Legitimate Businesses

Cartels can do a lot more than just compromise employees of a business. They can infiltrate, or even potentially take over, a legitimate business. The case of the MSC Gayane, a container ship owned by Mediterranean Shipping Company (MSC), the world's largest ocean carrier, provides a sobering example of this type of crime. In 2019, eight members of a criminal gang from Montenegro in central Europe joined the crew of the 11,600-container vessel.

The Gayane then made unplanned stops off the coast of South America, presumably to pick up cocaine from Latin American cartels. The ship was under surveillance. When it docked in Philadelphia a few days later, U.S. Customs and other law enforcement agencies boarded the ship and found almost 20 tons of cocaine, with a street value well over $1 billion hidden in shipping containers. According to prosecutors, four of the Montenegrin crew had taken jobs on the ship with the intention of engaging in cocaine smuggling.[6]

The Gayane case shows how deeply a cartel can penetrate even a major corporation. It also reveals, again, the potential for transnational cooperation between criminal gangs when the process suits everyone's mutual interests. Such dynamics could be a factor in space piracy and ground piracy affecting space commerce. For example, a cartel operating out of Asia might collaborate with a cartel in South America that can secure spacecraft landing facilities, and so forth.

No corporation will be immune to such infiltration. It may seem far-fetched, but cartel operatives could find work at major aerospace firms or space launch businesses. So positioned, the plants could facilitate space piracy from within the business.

An even more aggressive alternative might be for cartels just to buy a space business and run it for purposes of space crime at least part of the time. To understand the potential for such a move, it's important to grasp the changing nature of the space industry. When we think of private launch companies, for example, we picture SpaceX and a few other big ones. In reality, there are dozens of space launch startups, many of which are struggling to compete with SpaceX. Some industry experts predict a wave of bankruptcies,[7] which would make these companies easy targets for criminals intent on getting into space.

Professor Nathan Jones highlighted the difficulty of this situation from a law enforcement perspective. A cartel can become so enmeshed with a legitimate business that it is almost impossible to prosecute them. "If they're just buying and selling services on the market, what crime have they committed, even if the transactions are driven by threats?" he asked. This is a scenario that space piracy may favor: the activities of ostensibly legitimate businesses mask criminal conduct in space or on the ground.

Space-Based Money Laundering

Criminal cartels must launder their ill-gotten profits to avoid prosecution for tax evasion. To this end, they engage in a wide range of activities that mask the origin of illegal earnings, including buying and selling real estate, operating front companies, and so forth. Space presents a rich new venue for this process, with the ability to buy space-based assets and sell them as a way to "clean" dirty money. Alternatively, cartels might invest in space startups as a way to launder money, while also gaining access to space technology and talent for the purposes of space crime. We will cover this potentiality more in Chapter 9.

Why We Might Be Mistaken About Cartels as Space Pirates

Of course, we could be mistaken about the potential role of cartels in space piracy. One factor arguing against us is the current nature of piracy. As of today, most pirates come from the ranks of the desperate and dispossessed. Off the coast of Somalia, for example, piracy was ad hoc and performed by former fishermen who had lost their livelihoods. This does not match the cartel model.

Professor Correa-Cabrera, for one, thinks that space is an overly complicated environment for cartels. In her experience, cartels concentrate on activities that can make money with relative ease in the here and now. As space becomes more inexpensive and accessible, however, she is open to recalibrating her opinion.

What may emerge is a scenario wherein space piracy originates with small criminal groups, but is then coopted by cartels. This would align with the cartel model in many areas, where they establish a dominant role in a region and demand "tribute" from anyone committing criminal acts. A cartel could thus muscle in on space pirates if those pirates were operating in the cartel's territory.

Recruitment of Space Talent for Criminal Activities

Where would criminal groups, which don't likely employ too many space experts, find the talent they need to commit crime and piracy in space? One plausible answer is that they'll hire former space engineers and comparable talent in countries where such people are no longer in demand. Or they might compromise and control people who work in space industries and government space agencies through bribery.

Such scenarios are most probable in places where space talent is unemployed or underpaid. This means Russia and India at the current moment. Russia, whose space program and space economy have been badly impaired by international sanctions related to the Ukraine war, has many experienced space scientists either out of work or facing unemployment. As reported in the Moscow Times, work and life are growing increasingly difficult for space scientists. They are banned from international conferences as joint projects with NASA and EU are on hold.[8] These people are vulnerable to recruitment by criminal organizations.

India, a rising space power that just landed a vehicle on the moon for the astonishingly low sum of $35 million,[9] is another talent pool for potential space pirates. With obviously world-class space talent working in a developing economy, the potential for recruitment by criminal groups for space piracy seems high.

Takeaways

Who will be the space pirates? The world's criminal cartels are good candidates for this role, particularly if space piracy involves criminal activities on the ground. Criminal cartels are already expert in executing complex operations and infiltrating corporations and government entities. They could compromise space commerce businesses and proceed to engage in blackmail, theft, and disruption. Not everyone agrees with these predictions, however. Some experts believe that cartels will focus on "the here and now," and space is too far off for them to contemplate.

Notes

1. Vanda Felbab-Brown, "The Foreign Policies of the Sinaloa Cartel and CJNG – Part II: The Asia-Pacific," Brookings, August 5, 2022, https://www.brookings.edu/articles/the-foreign-policies-of-the-sinaloa-cartel-and-cjng-part-ii-the-asia-pacific.
2. Dave Hendricks, "Former Border Patrol Agent Sentenced to 10 Years in Prison for Protecting Cocaine Shipments," *Progress Times*, January 20, 2023, https://www.progresstimes.net/2023/01/20/former-border-patrol-agent-sentenced-to-10-years-in-prison-for-protecting-cocaine-shipments/#:~:text=A%20former%20Border%20Patrol%20agent,a%20hearing%20on%20Friday%20morning.
3. Luis Chaparro, "Mexican Cartel Bases used to Spy on Enemies, Dispatch Hitmen," Fox News, September 17, 2023, https://www.foxnews.com/world/mexican-cartel-bases-used-to-spy-on-enemies-dispatch-hitmen.
4. Ioan Grillo, "The Mystery of Mexico's Modern Pirates," Crash Out Media, May 25, 2023, https://www.crashoutmedia.com/p/the-mystery-of-mexicos-modern-pirates.

5. Benjamin Katz, "Criminal Gangs Exploit Security Gaps to Infiltrate Airports," *The Wall Street Journal*, October 3, 2023, https://www.wsj.com/business/airlines/criminal-gangs-exploit-security-gaps-to-infiltrate-airports-87576555.

6. Greg Miller, "How Shipping Giant MSC Reacted to Billion-Dollar Cocaine Bust," FreightWaves, April 5, 2021, https://www.freightwaves.com/news/how-container-giant-msc-reacted-to-billion-dollar-cocaine-bust.

7. Jeff Foust, "Small Launch Companies Struggle to Compete with SpaceX Rideshare Missions," *Space News*, October 18, 2023, https://spacenews.com/small-launch-companies-struggle-to-complete-with-spacex-rideshare-missions.

8. Vitaly Egorov, "Russia's Invasion has Thrown its Space Scientists out of Orbit," *The Moscow Times*, October 25, 2023, https://www.themoscowtimes.com/2023/10/25/russias-invasion-has-thrown-its-space-scientists-out-of-orbit-a82888.

9. Tripti Lahiri, "How India Pulled off its Frugal Moon Landing," *The Wall Street Journal*, August 25, 2023, https://www.wsj.com/world/india/indias-space-odyssey-to-pulling-off-its-frugal-moon-landing-41104972.

8

The Potential Impact of Space Piracy on National Security

The X-37B space plane at NASA's Kennedy Space Center in Florida after a 908-day mission in space
Source: Raymond Cassel/Shutterstock

S pace piracy, crime, and smuggling have the potential to affect the national security of the United States and other countries. This chapter explores what this might look like and who might suffer from these threats. It's not a simple issue, given the reliance that nearly every organ of national security has on space. Space war is not the same thing as space piracy, though at times, the two may resemble one another. Areas at risk include military intelligence and operations, diplomatic and geopolitical matters, and space industries that relate to national security.

The Stakes and Parameters of National Security

National security is a massive, thorny, and subjective zone of thought and action. The United States spends a good trillion dollars a year on it, and the issue is no less a priority for virtually every country on Earth. But, what is national security? This may seem like an overly basic question, but any meaningful analysis of space piracy's impact on national security should set itself in a clear context.

For our purposes, let's define national security first as a state of existence and then as a collection of activities. It is both. As a state of existence, a nation is secure when its people and economy are not under threat of harm or serious disruption by external or internal adversaries. Thus, the United States is relatively secure, while Israel, to name one example, is not. The former suffers from attacks here and there, but for the most part, Americans are safe from harm. In Israel, in contrast, civilians and the economy are under frequent, kinetic attacks from rockets and acts of terrorism.

As a collection of activities, national security comprises a broad set of institutions and processes aimed at delivering a state of security. These include the military, intelligence agencies, diplomacy, and national security industries. In the United States, we have the Department of Defense, but also the CIA, the NSA, NRO, and so forth. We have the State Department. And we have a huge constellation of defense contractors and related entities that build the products and deliver the services that help advance the cause of national security.

This complex array of organizations, processes, and technologies shoulders the burden of detecting threats against the United States, responding to them, or ideally preventing them from having an impact on the country. The U.S. national security apparatus does not operate in isolation either. We work closely with allies in gathering and analyzing intelligence. We seek to deter adversaries through alliances like NATO.

We need all the help we can get these days. Fortunately, the United States is not engaged in a major war. The country does find itself in conflicts that run the gamut from "hot," as exemplified by air missions in Syria, to cold conflicts like our adversarial relationship with Russia, to "shadow wars" such as the friction with China in the Pacific, and so forth. The United States is therefore constantly planning and budgeting for future conflicts. The planning process is based on intelligence gathering and analysis, among other factors.

As the national security apparatus goes about its work, it confronts threats that vary from serious and clear to subjective. A nuclear missile aimed at Washington, DC, is a clear national security threat. In contrast, industrial espionage targeting a tech company that makes AI software used by the U.S. Air Force is more of a gray area. National security organizations need to have a way of prioritizing such threats and allocating resources to mitigate them. The gray areas could loom large in space, driving a rethinking of national security policies and priorities.

National security pays attention to shifts in the balance of power around the world. Most of the time, luckily, the major powers are in stasis. Big changes or grandiose moves are rare, which is why, for example, Russia's invasion of Ukraine was such a shock to experienced national security hands. They were not expecting the war, but perhaps they should have, or could have, if they had been paying closer attention to intelligence.

The issue of casus belli, a cause for war, permeates national security. In the current era, it is not considered acceptable for countries to attack each other without reason. Casus belli might include actions like the blocking of a major port, the persecution of an ethnic minority, kinetic threats to sovereign territory, and the like.

Casus belli gets quite complicated in space. Is the blocking of a satellite signal an act of war? So far, the answer seems to be no. As is the case with cyberattacks, nonkinetic threats are not currently deemed to be acts of war. This could change in space.

In the NATO context, the acts that would trigger the treaty's collective self-defense clause are well defined. According to Article 5, "The Parties agree that an armed attack against one or more of them in Europe or North America shall be considered an attack against them all and consequently they agree that, if such an armed attack occurs, each of them, in exercise of the right of individual or collective self-defence recognized by Article 51 of the Charter of the United Nations, will assist the Party or Parties so attacked by taking forthwith, individually and in concert with the other Parties, such action as it deems necessary, including the use of armed force, to restore and maintain the security of the North Atlantic area."[1]

To date, there has not been an Article 5 event in NATO, though the Ukraine war is pushing at the edges of this consensus. The stakes are certainly high. A direct conflict between the United States and Russia or the United States and China could easily go nuclear, with untold catastrophes flowing from such an occurrence. Each country's respective national security organs, flawed as they may be, are wisely attuned to such risks and work to avoid all-out war. Space piracy has the potential to disrupt this status quo.

The Current Role of Space in National Security

Space permeates the national security landscape both directly and indirectly. As a direct element of national security, space systems like spy satellites provide intelligence data to the U.S. armed forces. Military satellites are also critical parts of command-and-control systems for weapons and military forces.

There are four basic types of satellites in use by the U.S. military:

- Communication satellites, which enable operational units to send and receive messages in real time. Many of these satellites are operated by corporations, not the military directly.
- Navigation satellites, which provide GPS navigation for military vehicles and personnel.
- Surveillance satellites, which monitor activities occurring on the ground.
- Reconnaissance satellites, which enable intelligence collection about U.S. adversaries.

The U.S. intelligence community relies on a host of satellite systems to gather data, including images and signals intelligence, about our adversaries. The National Reconnaissance Office (NRO) is one of the more prominent agencies involved in this work, but there are many others, some of which are so clandestine that their names and missions are not publicly known.

Space is also a theater of operations for kinetic weapons. Intercontinental ballistic missiles (ICBMs), for example, transit through space on their way to their targets. In that process, they interact with satellites for orientation and positioning. There are also Earth-to-space weapons, such as antisatellite missiles, and space-to-space weapons, which might use lasers to destroy satellites. Some of these are hypothetical or experimental, such as space-to-moon weapons. The Air Force's X-37 B pilotless space plane, to name one, is an advanced vehicle whose true purpose is secret.

Indirectly, the national security apparatus is dependent on space across several different domains. Satcom systems underpin ocean shipping, for example, which is essential to the U.S. economy, including defense industries. Satcom systems also anchor air traffic, which is essential for the movement of troops and goods related to national security. So, too, do national security entities rely on satellite infrastructure for part of their communications needs.

Disruptions to these space systems would, at a minimum, impair U.S. national security. An attack on a national security space asset could be an actual act of war, depending on how it was done and who perpetrated the attack. More on this in a moment.

More broadly, space assets underpin a great deal of America's economic activity. Freight transportation and logistics, air traffic, and many industrial sensors, such as those found in oil fields, depend on satellite connectivity. A disruption to that connectivity could negatively affect the economy. This could quickly evolve from being a nuisance to a national security issue if the public is faced with energy blackouts, shortages of food, and so forth.

A related risk has to do with the flow of information. Our adversaries have shown that they can and will flood the U.S. media with propaganda that favors their cause over the interests of the United States. They have done this, unfortunately, through mostly legitimate means, such as by deploying bots to disseminate opinions on social media. However, if they progress to hijacking communications satellites, our adversaries could inundate American media space with disinformation. This, too, is a potential national security issue.

Then, as the space economy becomes more important to the national economy, as is expected, threats to that space economy may also affect national security. For example, if the military requires rare-earth metals for its equipment and those metals come from the moon or asteroids, then a disruption in their delivery could affect national security.

The Current Space Players Affecting National Security

Until recently, each of the service branches of the U.S. military operated its own satellites. As of 2021 and 2022, the U.S. Navy and U.S. Army transferred operational control of their satellites to the new U.S. Space Force. This handoff was not a simple story, given that interservice friction is common. The service branches still use their satellites as they used to, but the day-to-day management of the satellites in orbit is up to the Space Force. They are the primary player when it comes to American national security and space.

What's important to note, however, is that private corporations are connected to national security space operations. Aerospace companies build the satellites. Technology companies create the computer hardware and software that is in those satellites. Private contractors are involved in maintaining the networks and other technologies that power the satellites. And private companies like SpaceX are often contracted to handle the launches.

The public–private partnerships that support national security space assets are significant because they broaden the attack surface of those space assets. Each vendor represents a potential path to sabotage and piracy. To date, this has not been an issue, but it is probable that we will see threats to national security space assets following an attack chain that involves corporate partners of national security organizations.

U.S.-Centric vs. Globalized Perspectives on Space and National Security

If you're an American, you are understandably interested in America's national security. You probably aren't losing sleep if Italy has a national security problem. Space, however, is inherently global in nature, even if the Air Force Space Command's 2019 report, *The Future of Space 2060 & Implications for U.S. Strategy*, did lead off with the declaration "For the last half century, the United States has been the dominant space power across the military, intelligence, science, exploration, and commercial domains."[2]

Yes, the United States is number one in space right now, and probably will be for some time to come. Things are changing, though, and an analysis of the risks of space crime and piracy would be more useful if it got away from a U.S.-centric view of space. The *Future of Space 2060* report, which is just one of many such policy documents, does get around to this perspective, arguing that "The U.S. must recognize that in 2060, space will be a major engine of national political, economic, and military power for whichever nations best organize and operate to exploit that potential."

It acknowledges that "The U.S. faces growing competition from allies, rivals, and adversaries for leadership in the exploration and exploitation of space." The report also points out that China is executing a long-term civil, commercial, and military strategy with the goal of economically developing cislunar space "with the explicit aim of displacing the U.S. as the leading space power." And it adds, "Other nations are developing similar national strategies."

Warning that "A failure to remain a leading space power will place U.S. national power at risk," the report recommends that the United States build a coalition to "promote and optimize the combined civil, military, and commercial exploitation of space to best serves the nation's interests."

Peter Garretson and Namrata Goswami, a former USAF colonel and space policy analyst, respectively, argue that space is on the verge of becoming a domain for "great power competition" in their book *Scramble for the Skies*. This exhaustive work analyzes many factors that will, in their view, drive conflict between the world's major economies as they strive to add space industry to their respective GDPs. As they see it, space will simply be another domain for the kind of economic competition we see on Earth.

Garretson and Goswami model several conflict scenarios in space. One view is that space will be a "monopolar" environment, in geopolitical terms. That is, a country like the United States or China will simply dominate space and either control or influence the exploitation of all space resources. Within this construct, they envision the potential for what they call "Protect of the Realm," wherein one country, like the United States, "provides exceptionally strong leadership, proactively extending a structure of liberal economy order into space."[3]

Alternatively, the monopolar "strategic configuration," as they call it, might feature a "Space Raj" where U.S.—China competition in space "continues to decay in both economies."[4] At the same time, India, a rising space power, experiences a surge of success in its space economy—leading to India becoming the largest economy on Earth. Based on that new position, minor space powers align with India for space economic activity and attendant national security matters.

Alternatively, they suggest that space could see a bipolar dynamic, where two powers, such as the United States and China each dominate a respective sector of space and engage in endless competition. There could also be multipolar scenarios, where no single power dominates and there is no clear "top two" competition, either.

Time will tell if Garretson and Goswami are correct, but their analysis is quite helpful in the present. It provides a structured way of thinking about space warfare and the potential for piracy to affect space war outcomes. Their approach is instructive, but it is limited by thinking that established Earth powers will determine the dynamics of space warfare. While it is impossible to predict, it certainly seems possible that other, current nonexistent players will affect the geopolitics of space.

Potential Bad Actors

Who will be the bad actors perpetrating space crimes and acts of piracy? At the risk of contradicting what we wrote in the previous chapter, where we identified criminal cartels as candidates for space piracy, when it comes to national security, it's likely that these organizations will not be acting alone.

Random acts of space piracy could affect national security, such as by blocking a satellite. However, if the intent of the space pirate is to cause harm to a country's national security, then that pirate will most likely not be perpetrating random crimes. Nor will he be working alone. Space piracy as a weapon against nation-states will almost certainly be conducted on behalf of other nation-states or comparable actors on the geopolitical scene.

Space piracy as a geopolitical threat could represent a rich variety of alliances between criminal organizations, nation-states, violent nonstate actors, and corporations. Possible examples include a nation-state using a criminal organization to harm an adversary while hiding behind that organization for the purpose of deniability. The Russian government's use of hacker gangs to damage American infrastructure is an ongoing example of this national security threat.

Alternatively, a rogue nation-state could commit acts of space piracy to enrich itself or disrupt other nation-states. North Korea's participation in financial crime is an example of this behavior today. The pirates in these scenarios would be acting as privateers, in essence, though perhaps without a public "Letter of Marque," as was the case in the old days.

Garretson and Goswami allow for this potential in *Scramble for the Skies*. They posit that "An aggressor state might issue licenses to harass or negate the operation of a rival state, and grant bounties for properties seized."[5]

They also outline a range of scenarios in space that could have an impact on wars being fought on Earth or in space. Some are criminal in nature. Others are more corporate but reflect the potential for trouble in space, where law enforcement and standard legal parameters are untested or nonexistent. They include the following[6]:

- **Claim disputes:** Situations where companies and/or countries could dispute a claim to a space resource, for example, the right to mine on an asteroid
- **Trafficking:** The illegal transportation of people or goods, for example, human trafficking or drug trafficking in space
- **Unsafe practices:** Where a company from country A acts in a way that harms the space activities of country B, for example, harmful space mining practices
- **Defection of citizens or facilities:** Where people from country A's space installation defect to country B's space facility, triggering a geopolitical confrontation
- **"Over-fishing":** Where country A violates an agreement not to extract more than its allotted share of a resource, for example, in space mining

The authors posit numerous other examples of how rogue actors, or criminals acting on behalf of nation-states, could cause serious conflict between space powers. They range from "lawfare," which involves negotiating agreements only to renege on them; claim jumping; limitations on innocent passage; the creation of "deniable dual use" military infrastructure; and more.[7]

As of the writing of this book, a situation is developing in the Red Sea that demonstrates the potential risks inherent in these scenarios. Houthi rebels in Yemen, who are backed by Iran, have been attacking Israeli shipping in the Red Sea. While not actual acts of piracy, the subjective and deniable nature of the attacks—they're probably ordered by Iran, but that can't be proven—make the situation difficult to address with standard military interventions.

The U.S. Navy is getting involved, because it's their job. As the Mission Statement of the U.S. Navy reads, "The United States is a maritime nation, and the U.S. Navy protects America at sea. Alongside our allies and partners, we defend freedom, preserve economic prosperity, and keep the seas open and free." As of now, there is no comparable force in space. The U.S. Space Force does not have such a mission statement today, but that may change in the future.

Perhaps more importantly, the attacks have the potential to explode into a far larger conflict. The crisis could result in the blockage of the Suez Canal, which would in turn paralyze a large portion of the global economy. All of a sudden, nation-states with no interest in the Houthi war or the Gaza war in Israel, would have reason to strike back, with serious implications that go far beyond the immediate, seemingly minor conflict at hand.

The same could occur in space. A small act of space piracy or crime, such as the blockage of a space facility, could lead to a wider blockage of much more valuable space commerce. This might then escalate into a confrontation between the space powers. A full-on war could ensue, spurred by an act that was not intended to cause that outcome. However, as practitioners of war know, once events start to happen, it's impossible to predict how the players will react and where things will lead.

Additionally, it might be wise to consider seemingly far-fetched but increasingly realistic threats of well-resourced private actors. Could a wealthy individual or a cash-rich criminal cartel create its own space weapons? That might sound like a silly question, but the potential of private citizens developing their own nuclear weapons is a real risk, according to a report from the Pentagon's Office of Net Assessment. As cited in a 2024 *Wall Street Journal* article,[8] the report claims it would take as little as one billion dollars and five years to build an atomic bomb...that's well within the reach of many individuals and nonstate actors. The same could be said for costly space weapons.

Necessary Digression: The Future of Human Habitation in Space

A discussion about space crime and national security needs to go on a tangent about human habitation in space. There are several reasons for this digression. For one thing, the question of whether the U.S. military needs to prepare human war fighters to travel to space to interdict pirates, or if robots will suffice, depends to some extent on who (if anyone) is living in space.

This question arises, or should arise, in our view, because of an even bigger question: can there be real sovereignty in space without human habitation? Put another way, can there be casus belli in space if human lives are not at stake? We don't have an answer for this question, but if history is any guide, issues of national security and war become pressing when populations are threatened. If space is inhabited only by machines, a threat to those machines may be of concern to their owners, but it might not trigger a national security response.

Professor von der Dunk of the University of Nebraska offered a perspective, which is that the use of armed force against American robots would not trigger the same legal consequences as such use of force against U.S. territory or human being. This is because, as he explained, the presence of American people in space would legally qualify as a kind of "representatives" of the United States to the extent that their activities directly would give rise to U.S. state responsibility and liability (per Articles VI and VII of the Outer Space Treaty). However, their presence does not carry any notion of territorial sovereignty over any part of a celestial body with it (per Article II of Outer Space Treaty). So, with this in mind, it makes sense that American machines in space also create no claims of national sovereignty.

Furthermore, the nature of space commerce and settlement will depend on whether human beings can live in space for extended periods of time. Fiction lets our imaginations run away from reality, at least from today's vantage point. We can easily imagine people living on large spaceships or on lunar or Mars colonies. But, will they? Knowledgeable people disagree.

One factor that's worth keeping in mind is that the vision of human habitation in space is driving public interest in the endeavor. The moment when a human being sets foot on Mars is going to be a lot more exciting than when a rover landed on the red planet. This is all the more reason to temper our excitement and try to be factual in assessing the national security aspects of space crime involving humans in space. After all, there may not be many humans in space.

Efforts are under way to put more people in space, however. As the International Space Station starts to go into the sunset, ambitious projects like Artemis aim to put people on the moon as soon as this year. In 2023, DARPA initiated a 10-year lunar architecture capability study known as LunA-10. Fourteen companies will be participating, each mapping out a future potential lunar infrastructure.[9]

Regarding Mars, which is the next logical place for humans to visit, the costs and complexity are exponentially greater than with the moon. The United States has many expensive and sophisticated robotic missions planned for Mars in the next decade. Meanwhile, billionaires like Elon Musk propose establishing Mars colonies for adventuresome human explorers. At this point, these visions seem likely to remain just that: visions and ideas that will take a very long time to come to fruition, if ever.

Cost aside, one of the biggest obstacles to human habitation on the moon or Mars is food. Assuming that it's too expensive to send rockets full of groceries to these remote sites, people in space will have to grow their own food. This is easier said than done, and the prospect of space starvation is not exactly what people like Musk have in mind when promoting their ideas. Even if this problem can be solved, the risk of criminals hijacking precious food in space is a risk to be considered. Latin American cartels take over food sectors, which might provide a preview of what's coming in space.

An arguably bigger problem is simple gravity. The moon has about one sixth of Earth's gravity, while Mars has 38%. As experience has shown even on relatively short space missions, zero gravity and low gravity are harmful to the human body. People lose muscle mass in low gravity. Fluids in the body flow in the wrong direction, causing swelling in the head that can affect eyesight, and more.[10]

It doesn't look too good, does it? Will people ever truly colonize outer space? We think the answer is yes, though the time frame is open to question. If you're looking at a two- to three-century time horizon, it almost guaranteed that humans will colonize space, for two reasons. For one thing, the food and gravitational issue can be solved with the right technology. And if there is money to be made by humans in outer space, they'll find a way to get there—and geopolitics, crime, smuggling, and piracy will surely follow them.

The Air Force report *The Future of Space 2060* offered a range of predictions on this subject. In their view, which is presumably based on research and work done by the Office of the Chief Scientist at the U.S. Air Force, there will be some human habitation in space by 2060. Their "lower-bound" projection is that "Humans have returned to the moon and landed on Mars, but there is no permanent presence at either."[11]

At the "upper bound," the report projects that "thousands" of individuals from many nations live and work from low Earth orbit (LEO), across the cislunar, moon, and Mars regions. This latter prediction is probably a bit optimistic for 37 years from now, but one never knows. After all, 37 years ago, the space shuttle Challenger blew up, making people wonder if space travel would ever advance at all in the future. And yet, here we are....

National Security Scenarios Potentially Affected by Piracy

National security planning often involves scenario-building exercises. What if? What if Russia jammed an American satellite? What if China launches a kinetic satellite weapon that could threaten satellites from other countries? What would the attack look like? How could the attack be prevented? How could it be detected? What is the correct response?

National security people and organizations devote countless hours to these questions. The main questions never change: What is the threat? How do we prevent it, detect it, and respond to it? The challenge with space is that we are in very novel territory. It's one thing to brainstorm how a new missile or armed group will affect national security. It's quite another to postulate on how national security events will play out in space, a domain where we have almost no warfighting experience and unreliable speculation about the conditions affecting national security in space in the future.

Nevertheless, a thorough approach to national security in space should take crime, smuggling, and piracy into account. These forces affect national security on Earth. So, too, should they be considered in space. We will attempt to do that here, starting with activities that are already occurring and then proceeding to more futuristic and unpredictable future scenarios.

Earth Warfare Affected by Space

Current wars here on planet Earth involve the use of many space systems. Whether the war is hot, cold, or shadowy, satellites are part of the story. And they are vulnerable to pirates who may or may not be working on behalf of the adversarial powers in question. A pirate group may, on its own initiative, hijack a national security satellite, either physically or virtually, and demand a ransom for its release. This act would affect the national security of the state that operates the satellite, but it wouldn't be a hostile act between two countries.

In contrast, a nation-state could induce a pirate group to commit an act of space piracy against a national security satellite operated by an adversary. The nation-state would have deniability. While it might be patently obvious to everyone who they ordered this act of piracy, they could say, "Oh no, it wasn't us. Piracy is against the law in our country, and we're doing everything we can to catch these criminals… (wink wink)."

Earth Warfare Fought in Space

We have not yet had a war fought between terrestrial powers in space, at least not one with kinetic attacks. That's not for a lack of weapons development, planning, and strategic policymaking, however. The most likely scenario for a war in space involves power A using antisatellite (ASAT) weapons to destroy a satellite from power B as a way to gain advantage on the ground. The goal might be to stop an adversary from gathering intelligence from space or from using satellites for GPS, command and control, and so forth.

The U.S. government created the Space Force in recognition of this type of threat. Announcing the creation of the new service branch, President Donald Trump said, "Space is the world's newest warfighting domain. Amid grave threats to our national security, American superiority in space is absolutely vital. We're leading, but we're not leading by enough, and very shortly, we'll be leading by a lot."[12]

At the time, many in the media and the public mocked Trump for setting policy based on some sort of *Star Wars* fantasy. Not only was this untrue and unfair, the negative media cycle covering the launch of the Space Force reveals a problem with space policy that could affect efforts to mitigate space piracy. Trump is a controversial and divisive figure. He elicits strong reactions from those who support and those who don't. The risk, as exemplified by this situation, is that politics will distort legitimate policy and strategic moves affecting space as a matter of national security.

What few people heard during the chorus of jeers for Trump's "stupid" Space Force decision was that the Space Force was really a consolidation and expansion of several existing Air Force space programs. Technically, the creation of the Space Force came about through a change in Title X of the U.S. Code, which defines the responsibilities and chains of command of each branch of the military. The change was the result of a new law, the U.S. Space Force Act, signed by President Trump.

The point to observe here is that government actions to deal with a problem like space crime take time and must necessarily involve matters of law and policy. The specific answer to the question "What will the government do to confront space crime and piracy?" is unknown. The general answer is well understood: it will involve no small amount of political wrangling leading to the creation of new laws and policies, or not, depending on the relative political power of the people and entities engaged in the process. This is not an optimal situation for addressing a crisis, but it's reality.

In the case of the Space Force, the political and legal process has led to the establishment of a branch of the armed forces with 8,600 personnel and responsibility to operate 77 spacecraft, including the GPS system, the X-37B spaceplanes, the U.S. Missile Warning System, military communications systems, the U.S. space surveillance network, and more. And following rules set out by the 1986 Goldwater–Nichols Act, the Space Force is on task to organize, train, and equip America's space forces, which are available to unified combatant commands in the military.

The operationalization of these mandates has begun. In August 2023, the Space Force announced the creation of its first unit dedicated to targeting adversaries' satellites. The 75th Intelligence, Surveillance, and Reconnaissance Squadron (ISRS) is part of Space Delta 7, a Space Force unit that provides intelligence on adversaries' space capabilities. The 75th ISRS will "analyze the capabilities of potential targets, locate and track these targets as well as participate in 'target engagement,' which presumably refers to destroying or disrupting adversary satellites, the ground stations that support them and transmissions sent between the two."[13]

The process of operationalization also included the establishment of the Space Warfighting Analysis Center (SWAC), which works on analyzing, modeling, and "war-gaming" space warfare scenarios. SWAC is analogous to the Army's Futures Command. It helps with operational concepts and force design, such as for missile warning and tracking, ground moving target indicator (GMTI); space data transport, position, navigation, and timing (PNT); tactical intelligence, surveillance, and reconnaissance (ISR); and space domain awareness (SDA).

This seems like a big workload, but the reality is that SWAC is just one of several entities, both inside and outside the government, that are dedicated to mapping out the future of warfare in space. Paul Szymanski's Space Strategies Center is one example. Szymanski has devoted nearly half a century to the study of space warfare. He has published dozens of papers on military space doctrine and served as an advisor to numerous branches of the government on the topic.

Szymanski's exhaustive analysis of space warfare, from current and future perspectives, identifies a number of issues that will confront anyone trying to think through the military's role in the mitigation of space crime and piracy. A 2022 article by Szymanski, "How to Win the Next Space War: An Assessment," published through the U.S. Air Force's Air University, highlights the depth of planning and strategic work required for success in a space war setting.

Szymanski lists no fewer than 36 "Top Rules to Fight and Win the Next Space War." These include factors like space situational awareness and weapons range, the pre-conflict positioning of space military assets, and the decisiveness of command.[14] Space situational awareness (SSA) and weapons range, which are fundamental requirements for any warfare in space, are about finding adversaries' pace assets and reaching them with space weapons. SSA may seem unnecessary. Can't we already see every objective orbiting Earth? Maybe not, especially if nation-states are constantly moving their space assets around and putting them in hard-to-reach orbits.

The military is on the case with SSA. One fascinating example is the concept of the Cislunar Highway Patrol System (CHiPs),[15] under development by the Air Force Research Laboratory. (Sorry, Poncho and John, of *CHPS* 1970s TV fame.) CHPS is envisioned as a system that generates SDA in the cislunar region, which sits between Earth and the moon. It's forward looking, because space traffic to the moon is still quite limited.

However, things may quickly change in this regard, with expected manned missions and potential industry and human activity on the lunar surface in the near future.

For Szymanski, intangible factors like political will and political consequences of space war are also essential to take into account when planning for space war. He points out that space warfare, at least at this time, will likely have no human casualties, so military and political leaders may feel emboldened to be aggressive—and he suggests that being the aggressor may be more rewarding than waiting to be attacked.

His insights and in-depth works are valuable in their own right, but they are also instructive for us if we want to imagine how the military will, or won't, get involved in combatting space crime and piracy. What he shows, indirectly, is that seemingly rapid actions taken by the military are often the result of very prolonged processes for studying and forming doctrines, followed by further long-lasting planning processes in the military. In parallel, there may be long cycles of weapons design and procurement.

Years, if not decades, can elapse between the time someone has an idea like "Let's build a stealth fighter jet that can take off from an aircraft carrier" and that plane's eventual deployment. Space will be no different, though it may in fact require even longer cycles of discussion, planning, procurement, force design, and training. The ideas that Szymanski and his colleagues are proposing may be 50 years away from being realized. But, he and others are wise to be thinking the issue through today. We need to take the same approach for space crime and piracy.

The Moon as a War Domain

As of 2024, the nations of Earth are abiding by the terms of the Outer Space Treaty and avoiding any direct, obvious militarization of the moon. However, efforts appear to be afoot to prepare to make the moon into a domain for warfare. Or, at the very least, major powers are increasingly looking at the moon as a domain as a military/strategic zone. DARPA, for instance, launched its Novel Orbital Moon Manufacturing, Materials, and Mass Efficient Design (*NOM4D) program in 2021.[16] This program is nominally about cultivating lunar industry. If so, then why is the Department of Defense leading it? The answer is that lunar industry and the militarization of the moon are closely linked objectives.

The Artemis Accords are similarly focused on peaceful activities on the moon, but it doesn't take a lot of imagination to think through how all the development of lunar colonization and industry might quickly translate into a need for military defense of such lunar assets. There are concerns that adversary nations will take the initiative to block any strategic advantage on the moon by the United States. For example, in 2023, the U.S.—China Economic and Security Review Commission presented a report to the U.S. Congress warning that "China seeks to control access to the moon for strategic aims."[17] In particular, the report expressed a concern that a Chinese satellite could block the Earth–moon L2 Lagrange point, which would allow China to fly to the far side of the moon and attack U.S. satellites in geosynchronous orbits. It's unknown as of this moment if the United States is preparing a countermeasure to this threat, but we suspect something like this is in the works, the terms of the Outer Space Treaty notwithstanding.

Private Armies and Other Nongovernmental Entities Affecting National Security in Space

The vastness and remoteness of space will have an impact on national security. There are long distances to travel to get from one point to another. And, even when the distance isn't so great, the technology required to move in space is costly and finite—to the point where traditional approaches to national security may not have much relevance.

This is not a novel situation, just one we haven't dealt with for a while. In the 1700s, for example, it took three months for a ship to sail from England to its colony in Australia. That created what might politely be referred to as a command-and-control problem for the British government. They had to rely on commanders to act on their own initiatives. Australia, as a political entity, had to act independently in many matters, even if it was not an independent country.

The Caribbean offered a similar dynamic that allowed piracy to thrive there. Official forces were slow to arrive, if they even did. Space is today's version of this kind of distance and loss of control. As the classic tagline for the sci-fi film *Alien* put it, "In space, no one can hear you scream." If a bad actor decides to commit a bad act in space, there might not be any force—military or law enforcement related—to stop him.

To protect themselves, corporations that do business in space may establish their own warfighting capabilities. This might involve hiring mercenaries or private defense contractors. Or it could mean space corporations creating their own fighting forces.

Such a move has precedent in the history of the British empire. In a bit of history that could foreshadow future developments in space, the British East India Company (EIC), which had been importing goods from India under royal charter since the 1600s, gained a further royal charter in the 1700s to field its own army. The EIC proceeded to engage in its own campaigns of conquest in India—controlling territory and signing treaties with local potentates at profitable advantage to the EIC.[18]

Space-based corporations might find themselves following a similar path to the EIC. With earthbound militaries unable or unwilling to defend their interests in space, they may resort to building their own military capabilities. Garretson and Goswami allude to this potentiality in *Scramble for the Skies*, noting, "An aggressor state might also provide a legal basis for a company to make its own law (as a company town or along the lines of European colonial joint stock companies) and act as its own armed force."[19]

Professor Margaret Sankey of the Air University also presented a scenario involving nonstate actors in space that could affect national security. "What would happen if a nonstate entity, such as a terror group, hijacked a satellite owned by an American corporation and used that satellite to dazzle satellites from, say, China? How would you go about apportioning blame and trying to figure out how to shut it down—especially if you're dealing with a company that operates transnationally?"

There is no simple answer here and the potential for conflict would depend on how eager the various players are for conflict. They might deescalate, which is usually the wise choice, or they could punish the corporation for its negligence by seizing assets or other aggressive moves.

The situation could also quickly get a lot more complicated when we consider that, in many cases, a privately owned satellite may quietly be under the control of a national government. Indeed, many space assets, even if privately owned, may have backdoors to government control. As Sankey put it, "The Chinese, for instance, have a tight relationship between companies and the government. Can we assume that any Chinese company satellite is also the property of their government or can be used by their government? Does that make them a target or a legitimate military asset in corporate clothing, so to speak?"

Alternatively, as the world has recently witnessed in the Ukraine war, a private citizen owning a private space company can make decisions that affect national security. As the CEO of SpaceX, which controls the Starlink constellation of communications satellites, Elon Musk refused to comply with an emergency request by the Ukrainian military to activate Starlink in Sevastopol, a region under Russian occupation.[20]

Musk described his decision by saying "The obvious intent being to sink most of the Russian fleet at anchor. If I had agreed to their request, then SpaceX would be explicitly complicit in a major act of war and conflict escalation."[21]

We can debate the wisdom of Musk's decision, but the unsettling fact remains: a private person, with no official connection to the U.S. national security establishment, has the power to control warfighting circumstances on the ground. As a space company, SpaceX has great potential to make similarly influential decisions regarding the space-related aspects of national security. SpaceX has many government contracts for launching U.S. satellites, so the United States could theoretically exert control over Musk's actions, but it remains to be seen if that will actually happen.

Controversial as Musk may be, however, it is reasonable to give him the benefit of the doubt as to his intentions. But, what if someone in his position wanted to exercise their space power for a bad purpose? What if a person of Musk's stature and power in space were aligned with an enemy of the United States? This is a question that the U.S. national security establishment should address itself to, if it isn't already doing so.

One risk is that spacefaring nations will outsource some of their military capabilities to private defense contractors in space. If they follow this path, they will be doing what is already a common practice on Earth. The U.S. military's use of Blackwater in the Iraq war showed the risk in this approach. The practice continues, however.

Companies like Constellis, which provide a broad range of military-style security services, are part of a growing industry. Constellis, which was recently taken over by a private equity management-led buyout, had $1.4 in revenue in 2023. The company operates in 50 countries. It is not a big leap of imagination that firms of this type could find eager customers in space.

Wars Between Space-Based Powers

While it may be somewhat distant in the future, it is possible that we will see the creation of entire space-based countries. Just as far-off colonies declared independence from Britain and France in the nineteenth and twentieth centuries, so too might space colonies eventually establish themselves as sovereign states.

Alternatively, we may see the rise of private companies that function as de facto sovereign states in space a la the British EIC. Obviously, this is pure speculation, but if history is any guide, large-scale human settlement in space will almost certainly lead to eventual political independence for those settlements. Space-based powers would also likely fight among themselves, just as any set of adversarial nation-states would on Earth. Piracy, smuggling, and crime would play a role in such conflicts.

It would be a mistake to assume that space-based powers would function like their Earthbound counterparts. For one thing, they would be springing to life in the tight technological confines of space, with all the rigid rules that come with surviving in a harsh environment. A Mars colony, for example, is not one where people can just do as they please. There would be strict control over what "citizens" were permitted to do—a sort of voluntary, reasonable fascism, if you will. Of course, that fascism could easily become unreasonable, and citizens would have little choice in the matter.

A Mars colony, or equivalent, would also be a "company town," at least in the beginning. People would be there to make money, perhaps through mining or some comparable industrial process. Their presence on Mars would be courtesy of a for-profit business, such as a major mining concern. The company would have its own private security forces.

The Mars colony might resemble the Saudi Arabian plan for a new city called Neom, a paradise of modernity in the desert—paid for with oil money. So, too, could a space colony be a paradise of modernity funded by space riches. The iron fist of monarchy would be the ruling authority in such a place.

With this kind of origin story, if the Mars colony declares itself to be a politically independent nation-state, it will probably not decide to be a democracy. It may be, as the German author Ernst Fraenkel described the Third Reich in the 1930s, a "dual state."

In Fraenkel's Nazi Germany, where he struggled to survive as a Jewish lawyer, there was a vast, invisible gulf between the official law of the land

and the actual law of the ruling, criminal Nazi Party.[22] A space nation might have a similar structure: a set of rules or laws on the books versus how things "really work," with hardline authoritarian controls, perhaps criminal nature, truly dictating what life is all about.

For Earth powers, dealing with space-based nations would be a serious challenge, considering the potential for them to function according to their own laws and norms. These norms would probably incorporate crime and piracy to an extent not seen in Earthly conflicts.

Evolving Space Alliances

As the costs of space travel come down, a growing and eclectic set of countries are getting involved in space for commercial and national security purposes. Some, like the United Arab Emirates (UAE) and Saudi Arabia, are relative newcomers. The UAE now runs four space research and development centers and even launched a probe that orbited Mars in 2021. This made them the first Arab country to reach Mars.[23]

Saudi Arabia, which is making many significant investments in space for scientific and military purposes, has established the Saudi Space Agency (SSA), a government agency dedicated to the kingdom's activities in space. Other Arab states, such as Bahrain, Kuwait, and Oman, have launched satellites in recent years.[24]

Others, like Japan, are adapting their space activities to meet new perceptions of threats and opportunities. In Japan's case, the most recent shift came in 2023 when the country announced that it would be changing the name of its Air Self-Defense Force to the Air and Space Self-Defense Force. According to Japan's Defense Minister, Minoru Kihara, the name change reflects a shift in strategy as potential threats from space become "more diverse, complex, and advanced."[25]

As new countries become spacefaring nations, the national security implications for all other countries in the world will change. It is not clear how things will change, but the realm of national security in space will become more complex as more countries have a presence in space.

The growth in spacefaring nations is already leading to some new potential alliances. For example, China is establishing collaborative and coinvestment relationships with the SSA and other Arab space entities.

These include China's apparent intent to develop its Tiangong space station in partnership with Saudi Arabia and the UAE.

What these novel and evolving partnerships will mean for U.S. national security is not clear at this time. However, the very fact that we're seeing such new pairings of nation-states around space is itself significant. The patterns of collaboration and alliances will likely grow more complex and unpredictable as space commerce grows and expands to other planets.

Takeaways

Space piracy has the potential to affect the national security of the United States and other countries, including those that have no space activities. Satellites are an integral part of military operations, intelligence, and industrial systems that support national security. Adversaries may resort to acts of piracy to threaten U.S. national security while denying direct involvement. Space piracy and national security represents a complicated topic, with many stakeholders and a huge range of risks that are difficult to assess at the present moment. However, what should be clear is that the national security apparatus would be wise to think through the potential for acts of space piracy to affect national security—and begin the long process of preparing for such threats. This thought process should take into account the potential for private actors like corporations to field their own paramilitary organizations, which may in turn function like armies in space.

Notes

1. "The North Atlantic Treaty," North Atlantic Treaty Organization, last modi fied October 19, 2023, https://www.nato.int/cps/en/natolive/ official_texts_17120.htm.
2. Air Force Space Command, "The Future of Space 2060 & Implications for U.S. Strategy," New Space Economy, September 5, 2019, https://newspace economy.ca/wp-content/uploads/2023/09/the-future-of-space-2060-3oct19-2.pdf.
3. Goswami and Garretson, *Scramble for the Skies*, 304.
4. Goswami and Garretson, 304.

5. Goswami and Garretson, 310.

6. Goswami and Garretson, 307–308.

7. Goswami and Garretson, 308.

8. Sharon Weinberger, "Could a Rogue Billionaire Make a Nuclear Weapon?" The Wall Street Journal, February 2, 2024, https://www.wsj.com/business/could-a-rogue-billionaire-make-a-nuclear-weapon-cd8bfde2.

9. Rachael Zisk, "DARPA Awards LunA-10 Moon Infrastructure Project Contracts," Payload, December 6, 2023, https://payloadspace.com/darpa-awards-luna-10-moon-infrastructure-project-contracts/?oly_enc_id=9807J1446578A7S.

10. Nathan Cranford and Jennifer Turner, "The Human Body in Human," NASA, February 2, 2021, https://www.nasa.gov/humans-in-space/the-human-body-in-space/#:~:text=Without%20the%20proper%20diet%20and,eyes%20and%20cause%20vision%20problems.

11. Air Force Space Command, "The Future of Space 2060."

12. Jim Garamone, "Trump Signs Law Establishing U.S. Space Force," U.S. Department of Defense News, December 20, 2019, https://www.defense.gov/News/News-Stories/Article/Article/2046035/trump-signs-law-establishing-us-space-force/#:~:text=Mark%20A.,20%2C%202019.

13. Brett Tingley, "U.S. Space Force Creates 1st Unit Dedicated to Targeting Adversary Satellites," Space, August 16, 2023, https://www.space.com/space-force-1st-targeting-squadron?utm_term=C84D30BD-6C07-4AB5-BAC1-0F7ADA07FC1F&utm_campaign=58E4DE65-C57F-4CD3-9A5A-609994E2C5A9&utm_medium=email&utm_content=3266C454-4F9E-47C1-B988-F458D3A868E0&utm_source=SmartBrief.

14. Paul S. Szymanski, "How to Win the Next Space War: An Assessment," Air University, April 4, 2022, https://www.airuniversity.af.edu/Wild-Blue-Yonder/Article-Display/Article/2981831/how-to-win-the-next-space-war-an-assessment.

15. David Buehler, "Cislunar Highway Patrol System (CHPS)," Air Force Research Laboratory, accessed August 9, 2024, https://www.afrl.af.mil/News/Photos/igphoto/2002556344/mediaid/4752579.

16. Brett Tingley, "DARPA Wants New Ideas for Space Weapons," Space, September 20, 2023, https://www.space.com/darpa-space-weapons-superiority-technologies.

17. Leonard David, "Why is there So Much Military Interest in the Moon?," Space, December 8, 2023, `https://www.space.com/military-moves-on-the-moon`.

18. Mark Cartwright, "The Armies of the East India Company," World History Encyclopedia, October 3, 2022, `https://www.worldhistory.org/article/2080/the-armies-of-the-east-india-company`.

19. Goswami and Garretson, *Scramble for the Skies*, 310.

20. Victoria Kim, "Elon Musk Acknowledges Withholding Satellite Service to Thwart Ukrainian Attack," *The New York Times*, September 8, 2023, `https://www.nytimes.com/2023/09/08/world/europe/elon-musk-starlink-ukraine.html`.

21. Kim, "Thwart Ukrainian Attack."

22. Ernst Fraenkel, *The Dual State: A Contribution to the Theory of Dictatorship*, trans. E a Shils and Edith Lowenstein. Lawbook Exchange, 2010, 266.

23. Oxford Business Group, "China Eyes Partnership with Gulf Countries for Space Exploration," Oil Price, August 27, 2023, `https://oilprice.com/Geopolitics/International/China-Eyes-Partnership-With-Gulf-Countries-For-Space-Exploration.html`.

24. Oxford Business Group, "Space Exploration."

25. Seth Robson and Hana Kusumoto, "Japan is Renaming its Air Force as Threats from Above Become More 'Complex'," Stars and Stripes, October 11, 2023, `https://www.stripes.com/theaters/asia_pacific/2023-10-11/japan-space-air-defense-force-11664863.html`.

9 Commercial Risks and Impacts of Space Crime and Piracy

A SpaceX Falcon 9 rocket
Source: SpaceX/Pexels.com

As of late 2023, international sanctions have prohibited Russia from selling crude oil for more than $60 per barrel. Yet, as Andrew Roth reported in *The Guardian*, Russia's "ghost," "shadow," or "dark" fleet,

"Nearly 500 ships, many of them old tankers with murky ownership and obscure insurers, could be playing an integral role in moving Russian crude to China and other ports in Asia, because of a G7 price cap meant to keep foreign-currency oil revenues out of the Kremlin's hands."[1] As Roth explained, the ships in this dark fleet move undetected by disabling their satellite tracking devices and other tricks.

So, despite sanctions and any number of international bodies whose job is to enforce the sanctions, Russia is selling billions of dollars' worth of oil right under the noses of everyone who objects, most strenuously, to this kind of thing. If that's how brazen sanctions evasion can be on Earth, imagine how bad things could get in space.

When it comes to assessing commercial risks due to space crime and piracy, the fundamental questions to ask are "What happens when there are no rules?" and "What will corporations and criminals, perhaps acting in concert, do when there is no law enforcement present?" We have many examples of bad behavior on Earth. In space, the outlook is not good, at least in the short term.

This chapter explores the potential impacts of crime, smuggling, and piracy on space commerce. We will explore such issues as corporate malfeasance in space, along with the potential for shell companies and false fronts to operate in space for malicious purposes. We'll examine the rich potential for money laundering in space, especially as driven by the use of cryptocurrency. This chapter will also look at the rapidly evolving category of insurance and risk management in space. We will further focus on the potential of organized bad actors, such as criminal cartels, to engage in corporate-type crimes in space.

Corporate Malfeasance

Space is likely to become a venue for corporate activities that are not illegal, per se, but still malicious and questionable. Garretson and Goswami make this point in *Scramble for the Skies*. They anticipate businesses in space engaging in anticompetitive tactics like blocking facilities and jumping claims.[2]

For Dr. Forrest Meyen, cofounder and chief strategy officer of the space robotics company, Lunar Outpost, there is a risk of piracy masquerading as corporate malfeasance. For example, based on his expertise in lunar exploration, Dr. Meyen pointed out that suitable landing sites on the south pole of the moon are finite in nature. As he put it, "Due to lighting conditions and surface roughness, coupled with landers that have an accuracy of 100 meters at best, a malicious actor could place an asset on the surface that effectively makes that landing site unusable—until a ransom is paid in some form."

The ransom could take the form of the target company agreeing to buy an overpriced product or accepting a low offer for its shares. A pirate entity could engage in this kind of extortion by acting through a front company. Alternatively, it could simply block the site and declare its pirate intentions out loud, demanding a ransom be paid in Bitcoin or the like.

Dr. Meyen believes that similar risks exist for certain space orbits. He said, "An orbit is a resource that is finite. Bad actors could place things to prevent other people from making useful economic good in a particular orbit. They park an unused satellite in the orbit and make it hazardous to fly or engage in useful economic activity of that orbit—until a ransom is paid, either in the style of piracy or through corporate blackmail."

Shell Companies and False Fronts in Space

Space pirates and comparable bad actors in space may disguise their activities by operating as seemingly legitimate businesses. This approach exists on Earth, so it is reasonable to assume it will occur in space. For example, as Professor Salvatore Mercogliano of Campbell University explained, Russia spent 2022 and 2023 "severing" portions of their main shipping companies "and creating little shell companies all around the world, in the UAE, in Oman, in Singapore, in a variety of countries all over the place." They're doing this not just to move but also for investments by the Russians outside of Russia.

Following this pattern in space, a malicious actor intent on smuggling stolen space-mined metals might resort to creating a front company to ship them from space to Earth. The front company might engage in legitimate shipping transactions as a cover, but its real business would be the hauling of space contraband. Or, a false front company could let itself be hired to transport a space cargo and then steal it. The potential for mischief in this regard is extensive, far worse than what could happen on Earth.

An alternative scenario for shell companies in space is what Garretson and Goswami call "deniable use military infrastructure" in space. An intelligence service, for example, could employ a front company to launch a satellite with an ostensible purpose like telecommunications, but which is in fact capable of being used for military communications. There are ample precedents for this practice on Earth. One notable example was the CIA's 1974 deal with the billionaire Howard Hughes to build the Glomar Explorer ship for the purpose of raising a sunken Soviet submarine.

We may already be seeing such a scenario playing out in space. Although the facts are not clear, it does seem that the controversy over the acquisition of the German satellite firm KLEO Connect by a Chinese corporation falls into the category of "deniable use military infrastructure." The story is long and complicated, but essentially, KLEO Connect is a German firm, but formed in 2017 as a venture with Lichtenstein's Trion Space and funding from Chinese investors. KLEO Connect's goal was to set up a global satellite network to compete with Elon Musk's Starlink and others.[3]

The company launched two satellites but has since fallen into a bitter dispute between the Chinese and German sides of the firm over control of the entity. A lawsuit ensued, with the matter ultimately reaching the level of the German economy ministry. In 2023, the ministry ruled that Shanghai Spacecom Satellite Technology (SSST), which held 53% of KLEO's stock could not acquire the entire company.[4] The reason? That's not officially known, but as Reuters reported at the time, "Germany has over the last year toughened its stance on China, with the government of Chancellor Olaf Scholz warning of the need to reduce its strategic dependencies on the Asian superpower."[5]

In this case, the German government was concerned that satellites serving the German military would be owned by a company described by knowledgeable observers, as "a military-connected, majority government-owned company specifically set up to invest in Western satellite projects (the PLA oversees all space activities in China)."[6] In other words, as German official watched Elon Musk refuse to switch on Starlink satellites over Crimea for fear of expanding a war he didn't approve of, they must have grown concerned that SSST could do the same to them at some time in the future.

Additionally, while SSST is alleged to have connections to the Chinese government, that means the company functions in some semblance of a structured military–intelligence–government framework. What if a company like SSST were owned by a criminal cartel operating through shell companies? Such a company could extort governments and militaries by threatening to turn off satellite networks.

Money Laundering in Space

Money laundering is a criminal activity that is likely to find space to be virgin terrain, ripe for exploitation. As with nearly everything else we've discussed so far, this is a guess, but it's probably a correct one. According to the United Nations, between 2% and 5% of the world's GDP is laundered globally. Criminals, terrorists, and others are therefore laundering somewhere between $800 billion and $2 trillion annually. It seems inevitable that this activity will extend to space as the space economy grows larger and becomes more integrated with the terrestrial financial system.

Assuming we are right about this, a brief digression into the nature of money laundering is in order. While there are many different ways to launder money, and an infinite variety of techniques within each of them, money laundering almost always follows a three-step process:

- **Placement:** The transferring of "dirty"/criminal money into a bank, shell company, or goods.
- **Layering:** Moving the dirty money or goods around so its origin is obscured.
- **Integration:** Converting the dirty money, now obscured, into cash or goods that the original launderer can declare as "clean" income for tax and reporting purposes.

With that in mind, we predict that space money laundering will follow two common methods of cleaning cash. One will be laundry through commercial transactions, such as the buying and selling of space cargoes. For insight into this first prediction, we spoke with Kenneth Rijock, a banking lawyer turned career money launderer, who eventually became a compliance officer and financial crime consultant. Rijock has a unique personal perspective into this issue.

In Rijock's view, money laundering in space will resemble conventional Earth-based practices, only it will occur in outer space. He said, "I think criminals will take the traditional money laundering formula of placement/layering/integration and apply it to cargo or some other goods of value, outside of regulated jurisdictions in space."

He added, "I'm sure you understand that there are probably people who stay up late at night trying to figure out how to bring law to space. And they may find some sort of way to do so, but that won't stop the money launderers because they'll just transplant their method aloft. They'll buy goods with dirty money, like minerals from asteroids for cash. That will be the placement stage. Then, they will arrange for those minerals to be purchased by a company they control on Earth…which is the layering process. When that firm sells those minerals, that's integration. They get 'legitimate' taxable income from the sale of space minerals. Just like Al Capone and his chain of laundromats in old Chicago."

The other mode of money laundering that's probable in space involves shell companies and the buying and selling of space assets. This is a widespread practice on Earth right now, often involving real estate and other assets that are subjective in value, such as artworks. According to LexisNexis, "Shell companies and trusts are effective tools used in money laundering schemes, as they offer an intricate façade to disguise illicit funds."[7]

With a shell company, a money launderer can throw up a smokescreen that enables the transfer of funds between accounts—the illusion of legitimate transactions. Then, as LexisNexis explained, "By intertwining multiple shell companies, the complexity of the financial web grows, making it increasingly difficult for authorities to trace the money back to its original source."

Luxury high-rise apartments in New York city are a prime example of this technique. Opaque, multilayered shell companies with foreign ownership buy and sell Manhattan condos worth tens or even hundreds of millions of dollars. It is an open secret that many of these transactions are masking money laundering.

Space will almost certainly be a venue for this kind of activity, according to Professor Mercogliano, the maritime piracy expert. He said, "I believe that money laundering operations will include the buying of expensive space assets, because it's an easy way to launder money—easier than, let's say, a shrimping company or siphoning oil off a Panamax ship. For the actual purchase, they will create front companies or investment firms that will buy legitimate space entities—and then sell them to other companies who are colluding with them, either knowingly or not."

The potential for shell company-driven money laundering in space offers the beginning of an explanation for why small countries like Luxembourg have taken such an active interest in space commerce by becoming a full member of the European Space Agency (ESA).[8] At a glance, Luxembourg, a tiny landlocked country, seems like an unlikely candidate to be a space power.

Yes, the country's business elite may legitimately want to get a head start on a growing, high-value industry. However, a better explanation is that the country, known for harboring questionable financial entities, sees itself as a vessel for asset-based space money laundering.

With the caveat that this may turn out to be unfair insinuation, recent reporting suggests that space-based money laundering could be in Luxembourg's future. The OpenLux investigation, conducted by reporters from *Le Monde*, *Le Soir*, *The Miami Herald*, and *Sueddeutsche Zeitung*, examined millions of documents related to 260,000 companies linked to Luxembourg's €4.5 trillion ($5.4 trillion) investment funds sector between 1955 and 2020.

The investigation concluded that the country's investment fund industry is "a financial 'black box' that helps people launder illicit money and avoid tax."[9] Will Luxembourg's "black-box" investment fund sector open itself up to space-based money laundering? We'll have to wait and see…assuming we can see what's happening there.

Mercenaries and Corporate/Colonial Exploits in Space

The size and remoteness of space create opportunities for nonstate actors to engage in a blend of commerce and military activities without undue concern about consequences. Yes, at some point, we will have well-defined areas of national sovereignty in the key commercial regions of space. In the near term, however, which could be measured in centuries, we are going to be dealing with a lot of subjective power vacuums that reward criminals as well as corporate entities that are in the business of fighting wars.

Ironically, it may be a desire to defend space-based industry against pirates and other crimes that leads to devastating unintended consequences for space corporations and their home countries. It could look like this: a space-based mining company hires a private military contractor (PMC) to protect its space mining outposts. At some point, the PMC decides, for reasons of its own, that it would rather control the mining assets than guard it. Armed and unchallenged, they proceed to take over the operation and begin to sell the mining output on the black market back on Earth.

History is full of such tales, going as far back as the ninth century, when the ruler of the Abbasid Caliphate hired Turkish mercenaries, but failed to control them. After a series of revolts by the Turkish soldiers, the actual Caliph became a puppet of mercenary strong men. These historical episodes all work from a central premise: if you're too weak to defend yourself, hiring mercenaries to fight your battles may result in those mercenaries exploiting your weakness in the end.

This could happen in space. A spacefaring power hires mercenaries or some other form of private security contractor to protect its space interests...and watches helplessly from Earth as the mercenaries declare themselves sovereign over the space assets they were hired to protect. The remoteness of space and the high barriers to space travel could easily lead to such a situation.

A different space mercenary scenario with bad, unexpected outcomes would be one where a legitimate nation-state uses mercenaries to mask its geopolitical moves—creating deniability for aggression. This sort of thing has played out many times in history, and it's arguably happening right now with Russia's Wagner Group in Africa.

The Wagner Group's activities in Africa represent a sprawling, complex topic with more than a few opaque segments where speculation is required. Reporting suggests, however, that the Wagner Group has taken an active role in intervening in countries like Mali and Libya, often in the aftermath of a coup or other destabilizing events. The group provides armed support for political factions while investing/exploiting the country's economy at the same time.

In Mali in 2021, for example, the Wagner Group made a deal with a military junta that had recently taken over the country. According to an analysis by IRSEM, a research branch of the French military, Russia did not deny this, but comments by Russian Foreign Minister Sergey Lavrov at the United Nations are revealing. He confirmed that talks were occurring between the government of Mali and a Russian PMC. He told the General Assembly, "Insofar as the Malian authorities consider their capacities insufficient without external support and this external support is decreasing, they have turned to a Russian private military company." Vladimir Putin later said that Russian PMCs "could be present in the country to participate in operations to secure gold deposits."[10]

The IRSEM report is titled "A Foreign Policy by Proxies?" which aptly describes what's actually going on here. The Russian government can casually remark that a Russian PMC is active in Mali, as if it were simply some sort of business deal. What's more likely is that Wagner Group is executing a specific strategic plan for the region on behalf of the Russian government—but the arrangement appears to be occurring at arm's length.

We may see such moves in space, where PMCs or the equivalent go about implementing aggressive but economically valuable geopolitical moves on behalf of nation-states. The process provides deniability and reduces the chance for an act of war in space to translate into real warfare in space or on Earth.

The Wagner Group's financial picture is beyond cloudy, but it's evident that the group has significant investor backing. Its operations are deep and expensive, so someone is paying for it. That might be the Russian government, but it could also be any number of super wealthy individuals or companies that want a piece of Mali's gold mining industry, to name one of many business ventures being pursued by the PMC at this time.

Space mercenaries could similarly obfuscate illegal resource grabs by private entities. Wealthy people or rich but obscure shell companies might find it advantageous to fund mercenaries that do the dirty work of seizing space assets on their behalf. Experiences on Earth, such as the Wagner Group's, show how it might work in space.

Space Banking, Piracy, and Cryptocurrency

It's early days for banking in space, but the idea is gaining traction, at least in research circles. A 2016 paper by Bruce Cahan from Stanford and Irmboe Marboe from the University of Vienna, "Outer Frontiers of Banking: Financing Space Explorers and Safeguarding Terrestrial Finance,"[11] sets out the thesis that the emerging space economy will require specialized banking capabilities.

In the 2016 paper, the authors note that banking and insurance capacity were "preconditions to seafaring explorers and mining operators." Likewise, they state "Space needs its bankers and insurers to take a holistic view of the space economy to attract and invest capital across a broad range of private companies and their inter-related and interdependent projects, project teams, investors, and technological achievements."

In 2016, the authors mostly focused on terrestrial banks doing business with terrestrial space firms. However, as space banking expertise grows in parallel with increasingly ambitious space commerce, it will only be a matter of time until hybrid or exclusive space banks start to operate in space. In 2018, Bruce Cahan and his coauthors described a Space Commodities Exchange for trading goods and financial derivatives between Earth and space and wholly within space domains (LEO, MEO, GEO, cis-lunar, and beyond).[12] We'll return to this idea in Chapter 10.

Would terrestrial financial regulators be sufficient to regulate such banks, commodities exchanges, and other financial market intermediaries? Space could become the new Cayman Islands, a place that indulges customers in questionable banking transactions. They could be aided by space-based law firms and the ability to incorporate business entities based in space, with its flimsy legal system.

For example, how would space banks avoid being on the front lines of space-based money laundering and related crimes? What types of data, transparency, trust, and other processes would keep such space financial mechanisms from becoming sites for deliberately or unwittingly storing, laundering, transmitting, and processing illegally gotten funds from space smuggling and the like? Think about every space-based science fiction story from *Star Trek* to *Star Wars* and *Dune*—and ask whether know your customer (KYC) even assurable in the space commerce era?

Regulated financial institutions like banks, commodities exchanges, and securities markets may be essential to grow a responsible space economy. As shown on Earth in recent years, legitimate banks are unnecessary for money laundering in space. The ability for bad actors to move funds around using cryptocurrencies presents an attractive alternative to placing/layering/integrating funds through banks. This practice is already standard for cartels on Earth, as the arrest of Ignacio Santoyo, a Mexican human trafficker, revealed in 2019. His alleged use of Bitcoin was what got him caught. Santiago Nieto, head of the Mexican finance ministry's financial intelligence unit (UIF), remarked, "There's a transition to committing crimes in cyberspace, like acquiring cryptocurrencies to launder money."[13]

Notwithstanding the costs for launch and other operational considerations, space is an especially rich resource for blockchain/digital ledger, artificial intelligence, and other energy-intensive computer applications, given the access to solar energy and cold temperatures. Computers are in space, ergo, cryptocurrencies will soon follow. The process is well underway, with playful ideas like erecting a statue on the moon that contains a non-fungible token (NFT)[14] to more serious announcements like the launch of Cryptosat, which intends to create a "Space Wallet" so users can spend and receive cryptocurrencies in space.[15]

The venerable investment banking firm of J.P. Morgan has joined the fray. In 2021, J.P. Morgan "tested the world's first bank-led tokenized value transfer in space, executed via smart contracts on a blockchain network established between satellites orbiting the earth."[16] The firm is also pioneering blockchain-based satellite-to-satellite through its SpaceBridge project.[17]

It's hard to predict if and how cryptocurrencies will factor into space-based money laundering. Criminals will probably stay away from platforms like SpaceBridge sponsored by regulated terrestrial institutions such as J.P. Morgan. Yet, if money launderers are hiding behind front companies that appear to be regular businesses, they may indeed use space blockchains and wallets created by actors outside of the regulated marketplaces on Earth or elsewhere.

Insurance and Risk Management in Space

The potential for piracy and crime in space is beginning to affect how the insurance industry views space as a venue for doing business. For insights into the shifts in thinking, we spoke with Denis Bensoussan, head of space at Beazley, a provider of risk products in the United Kingdom. For an experienced insurance professional like Bensoussan, space appears to mimic the pattern the industry has witnessed with maritime and aviation insurance.

As he pointed out, insurance as we know it today originated with marine insurance, first offered by Lloyds of London about 300 years ago. Lloyds insured cargoes, such as gold and spices, which were going back and forth to the new world. He suspects space will be the same, once there are valuable cargoes to be transported.

For now, insurance for space covers the space vehicle, such as the satellite or rocket, as well liability, in case the rocket crashes. "There hasn't been interest in insuring space cargoes," Bensoussan explained, "because space cargoes aren't worth much. A rocket to the ISS carries food one way and trash on the return trip. Why insure that?"

This will likely change, in his view, as the space economy develops. What's important to understand, he commented, is that insurance policies have to cover a specific area of loss if the customer wants to be insured. This may sound obvious, but space companies that "get insurance" may not be covered for acts of piracy. There has to be distinct piracy and war coverage. In maritime, for example, "situations of piracy and war or criminal scenarios were excluded from standard policies," Bensoussan shared. "So basically, your cargo was going to be covered if there is a sort of a storm or if the boat sank or if the cargo gets lost because transit takes too long, or was handled badly during docking operation or something like that. That was covered. But, what was excluded was piracy."

Space insurers will need to develop products that don't exclude piracy and acts of war if they want to cover such scenarios. The insurers may follow a path taken by aviation insurance carriers in the wake of the 9/11 attack. Prior to that, acts of war were not covered, but afterward, aviation firms had the option of carrying insurance against war and terrorism.

One reason Bensoussan thinks that space insurers will develop policies that cover piracy and acts of war is the relationship between insurance and the financing of business ventures. In maritime trade, for example, commercial ventures require financing for the shipments. Lenders, in turn, will not extend loans for shipments that are not covered by insurance. As piracy picked up off the coast of Somalia, lenders required coverage for acts of piracy. This will likely occur in space, too.

Smuggling and Dark Fleets in Space

Smuggling, along with the phenomenon of "dark fleets," represents another area of concern regarding the commercial impacts of criminal activity in space. Best exemplified today by Russia's trade in sanctioned oil, but by no means a new idea, dark fleets show how persistent and dedicated bad actors can flout international law.

Michelle Wiese Bockmann, principal analyst at Lloyd's List Intelligence, offered insights into how the dark fleet operates. She frames the issue by first describing the three layers of maritime regulations and conventions. At the top, there is the UN's International Maritime Organization (IMO), which has a whole series of conventions and regulations that apply to the organization's 180-plus members. Then, there are the rules for the flag countries where a ship is registered, for example, Panama. Port state control is also a factor, with rules and regulations affecting ships based on where they dock and disgorge their cargoes. Finally, there are classification societies, which set rules for structural integrity of ships and certify if they are seaworthy.

With the dark fleet, according to Wiese Bockmann, what's going on could be called "regulatory arbitrage." She says, "Ship owners look and think, 'I want to be outside Western jurisdictions. I'm going to choose a flag that has no technical or maritime expertise. Perhaps Gabon is a good flag state for that.' Then I'm going to pick a recognized organization that doesn't really care if I do a special survey of the ship every five years, or they won't look too closely. And then, I'm going to get my blue card, which is my insurance card. I'm going to get one that is recognized by flag (that I'm registered with), but doesn't look to see whether or not I have the right kind of liability coverage."

Before you know it, as Wiese Bockmann explains, you have a ship that is the equivalent of an old banger: "You know, really a rust bucket that's hurtling down the highway with no insurance, in a terrible dilapidated state. And everyone's going, 'What's going on with this?' But, it's perfectly legal. That's how we have the dark fleet."

The dark fleet, now numbering in hundreds of ships, is moving oil from Russia to customers all over the world. These ships may off-load their oil cargoes offshore, ship to ship, away from port control. They turn off their satellite transponders, so they are difficult to track. They're an insurance and environmental catastrophe waiting to happen, but as Wiese Bockmann suggests, it's all legal, and who will stop them?

In her view, space represents a comparable arena for dark fleet types of activity. "It's speculation," she cautions, "but you potentially have a lot of the same forces at work: commodities that may be under sanction, at least on Earth, but buyers and sellers who want to conduct business anyway. And, as with oil, you will likely have rogue traders who know how to work around the official system. The regulatory landscape is almost nonexistent at this point, so breaking rules may not even be a relevant issue. The major difference between maritime and space dark fleets is that space cargo transport is new, or even hypothetical, at this point. It will be quite a while before we have old spacecraft that no one wants to track any longer. But, it will probably happen."

Takeaways

As space commerce grows, we are likely to witness a range of criminal and semicriminal activities that threaten both the commercial sector itself and nation-states that engage in space commerce. The issue may be as simple as corporate malfeasance, which involves actions like claim jumping or blocking of limited resources, like orbits, to extort corporate advantage from the victims. This is not illegal, but it's not right, either, and it could cause friction in space industries.

Shell companies or false fronts also present a legal but questionable source of harmful corporate activity in space, as well as a basis for money laundering. Indeed, money laundering experts see space as a rich venue for this illegal practice. Space pirates and smugglers will likely have many options to launder their ill-gotten funds. Terrestrial criminals may also avail themselves of space-based money laundering opportunities.

Mercenaries and comparable private security entities in space could be a necessity but also a threat. As history has shown, leaders of private armies sometimes become sovereign leaders, overthrowing their clients. The remoteness of space makes this probable.

Banking and insurance will be necessary in space, but it's not clear how such companies will actually function. Terrestrial laws may apply, but it's not a good assumption to make. Many factors in space argue for a separate set of regulations to cover this unique arena for commercial transactions.

Notes

1. Andrew Roth, "The 'Dark Fleet' of Tankers Shipping Russian Oil in the Shadows," *The Guardian*, August 19, 2023, https://www.theguardian.com/business/2023/aug/19/the-dark-fleet-of-tankers-shipping-russian-oil-in-the-shadows.

2. Goswami and Garretson, Scramble for the Skies, 309.

3. Comms, "Dispute about Huge Space Project from Liechtenstein," Rivada, May 16, 2022, https://www.rivada.com/dispute-about-huge-space-project-from-liechtenstein.

4. Reuters, "Exclusive-Berlin Blocks Complete Takeover of Satellite Startup by Chinese Firm – Sources," U.S. News, September 13, 2023, https://www.usnews.com/news/technology/articles/2023-09-13/exclusive-german-government-forbids-complete-takeover-of-satellite-startup-by-chinese-firm.

5. Andreas Rinke, "Exclusive: Berlin Blocks Complete Takeover of Satellite Startup by Chinese Firm," Reuters, September 13, 2023, https://www.reuters.com/markets/deals/german-government-forbids-complete-takeover-satellite-startup-by-chinese-firm-2023-09-13.

6. Glenn Chafetz and Xavier Ortiz, "China, Lawfare, and the Contest for Control of Low Earth Orbit," The Diplomat, August 10, 2023, https://thediplomat.com/2023/08/china-lawfare-and-the-contest-for-control-of-low-earth-orbit.

7. "Examples of Money Laundering Techniques," Lexis Nexis, May 4, 2023, https://www.lexisnexis.com/blogs/gb/b/compliance-risk-due-diligence/posts/examples-money-laundering.

8. Marc Serres, "How Luxembourg Becomes Europe's Commercial Space Exploration Hub," Cairn, May 2019, https://www.cairn.info/revue-realites-industrielles-2019-2-page-69.htm.

9. Huw Jones, "Luxembourg Fund Industry is a \$5.4 Trillion 'Black Box', Investigation Says," Reuters, February 8, 2021, https://www.reuters.com/article/idUSKBN2A81NO.

10. Maxime Audinet and Emmanuel Dreyfus, "A Foreign Policy by Proxies? The Two Sides of Russia's Presence in Mali," IRSEM, September 2022, https://www.irsem.fr/media/report-irsem-97-russia-mali-en.pdf.

11. Bruce B. Cahan, Irmgard Marboe, and Henning Roedel, "Outer Frontiers of Banking: Financing Space Explorers and Safeguarding Terrestrial Finance," New Space 4, no. 4 (December 2016): 253–268, https://doi.org/10.1089/space.2016.0010.

12. Bruce B. Cahan, Pittman, R. Bruce Pittman, Sarah Cooper, and John Cumbers, "Space Commodities Futures Trading Exchange: Adapting Terrestrial Market Mechanisms to Grow a Sustainable Space Economy" New Space 6, no. 3 (September 2018): 211–226, https://doi.org/10.1089/space.2017.0047.

13. Diego Ore, "Latin American Crime Cartels Turn to Cryptocurrencies for Money Laundering," Reuters, December 9, 2020, `https://www.reuters.com/article/idUSKBN28I1K2`.

14. Ben Weiss, "Blockchain to the Moon! Crypto Companies' Outer Space Plans are Less Silly than You Think," Fortune Crypto, July 4, 2023, `https://fortune.com/crypto/2023/07/04/blockchain-space-blockstream-spacechain-filecoin-cryptosat-satellites/amp`.

15. Weiss, "Blockchain to the Moon."

16. "Onyx by J.P. Morgan Launches Blockchain in Space," JP Morgan Chase & Co, accessed August 10, 2024, `https://www.jpmorgan.com/technology/news/blockchain-in-space`.

17. "SpaceBridge: Pioneering Payments in Space," Onyx, accessed August 10, 2024, `https://www.jpmorgan.com/onyx/payments-in-space`.

10 Policy Recommendations and Countermeasures to Mitigate the Risk of Space Piracy

U.S. Space Force Guardian experiences zero gravity in a training flight conducted as part of the Space Force's demanding "Azimuth" aerospace missions training program.
Source: U.S. Space Force

If you've stayed with us this far, you probably get that space crime and piracy are potentially serious threats to the coming space economy. As we've discussed, the specifics of what these problems will look like are impossible to know, but they are possible to guess with a degree of accuracy. Trouble is surely on the way. The question that should be in the tops of our minds, then, is "What can be done about this?" That's what this chapter is about.

We will start with a quick overview of the current laws, policy frameworks, and institutions that are logically in line to deal with issues of space crime and piracy. It is our view that, in their current forms, these laws, policies, and frameworks are not up to the challenge of mitigating space crime and piracy. That is not anyone's fault. The need simply hasn't arisen yet, but now is the time to focus on it.

Additionally, the current stakeholder group is incomplete, in our view. In this chapter, we will explore the entities that should have a role in addressing space crime and piracy. From there, we will discuss the critical need to agree on principles as a first step in solving the problem. It is only from a shared set of values that any coherent or effective mechanism for combatting space crime and piracy will emerge.

The rest of the chapter gets into specific recommendations regarding the potential to adapt maritime law to space, the need for more robust space governance and clarity on sovereignty, space law and law enforcement, the roles of military and intelligence establishments, and the need for financial and insurance industry players to get involved in the dialogue.

Restating the Problem: The Limits to Current Frameworks and Institutions

As we discussed in Chapter 6, the legal and treaty landscape for combatting crime, smuggling, and piracy in space is not empty. We have the Outer Space Treaty and several other important international agreements regarding national conduct in space. Law in space is uneven and subjective, but there are laws on Earth that apply to what people do in space, up to a point. National militaries and their counterparts in intelligence are devising strategies and operational programs with the potential to counteract criminal behavior in space.

So, what's the problem? Well, maybe we don't even truly understand what the problem is at this moment. That itself is a problem. As Donald Rumsfeld might have put it, we have "known unknowns" and "unknown unknowns." However, what is possible to see today is that there is no coherent legal or treaty framework that can effectively deal with space crime and piracy. Jurisdiction, law enforcement, courts of law, and sovereignty are subjective in space, and that's being generous. Nor is there a coherent military or intelligence solution, at least one that is publicly known.

The essential challenge is that dealing with space crime and piracy will have to be a multithreaded effort. Commercial interests are at stake and require commercial solutions. However, national security is at risk, too. That must also be factored into a solution. Law enforcement processes are part of the picture. An act of piracy might affect a business, but it could also represent a deniable attack by a nation-state. No single sphere of policy can handle all of these dimensions at once.

Practicalities also matter. For instance, speed of response is critical. If there is an incident of space piracy, who will respond, how, and how quickly? Who will bear the cost of interdiction? Even with robust laws on the books, if rapid action is required and none is forthcoming, that's going to be a big problem.

What the world now has, at best, are siloes of policy and response mechanisms to prevent, detect, and respond to incidents of space crime and piracy. No one has an overview of the situation. Roles and collaborative structures barely exist. This is not an ideal situation, but plenty of time exists to focus on solutions.

Whose Problem Is This to Solve?

Assuming that space crime, smuggling, and piracy become problems, whose problem is it to solve? As the complexities and costs of dealing with ocean piracy in recent years have demonstrated, it can be a difficult question to answer. Everyone and no one, it seems, is responsible for dealing with piracy. This may turn out to be the case with space, but there is an opportunity to do better.

Today, we have silos of responsibility and interest when it comes to dealing with potential crime in space. Each silo, such as national governments, corporations, law enforcement, military, and so forth, has a part to play in mitigating space crime. However, it's not reasonable to expect a single silo to shoulder the entire burden. Indeed, it would be exceptionally difficult for any one of them to do that even if they wanted to.

Rather, responsibility for space crime and piracy must be collective in nature. It may be optimal, therefore, to create some central governing body that can establish norms and rules—and then bring stakeholders together to forge consensus, agreements, and action steps. Such a body might be part of the United Nations (UN), such as the UN Office of Outer Space Affairs (UNOOSA), but that should not be a foregone conclusion.

The UN is already set up to facilitate agreements regarding outer space, but the organization is not well equipped to deal with enforcement and the kind of operational requirements that an effective antipiracy regime would need in space. What might be optimal would be for the necessary initial dialogues and high-level agreements to occur in a UN setting but then migrate to a new body that has the means and interest in taking concrete actions.

Alternatively, and perhaps more realistically, the anticrime and antipiracy rules and norms affecting space will be part of a broader scheme of treaties and laws that affect overall national conduct in space. For example, if there is a consensus, followed by international agreements, regarding the nature of national sovereignty in space, then it might make sense for measures dealing with crime and piracy to be part of that consensus.

Start with Principles

Assuming such a body comes into being, how should they approach the thorny and subjective challenges of establishing a lawful commercial and geopolitical ecosystem in space? History offers some good procedural precedents. If you look at how the Outer Space Treaty came into being, or even the Geneva Conventions, you will see that the final product came as the result of lengthy and thoughtful deliberations by bodies representing multiple nations. Something similar could work for space crime.

For insight into this unusual challenge, we spoke with Annie Pforzheimer, whose 30-year career as a foreign service officer at the U.S. Department of State included serving as Acting Deputy Assistant Secretary for Afghanistan. Pforzheimer has a deep appreciation for the potential for international agreements to make a difference, but also a strong sense of how fragile they can be when conflict puts stress on them.

In her view, space is a "commons," similar to the sea, but with some notable differences. Unlike the sea, for which leaders could draw on millennia of commercial and military tradition, space is untested. There are few rules, but Pforzheimer cautions against starting the process by establishing rules. "If you come up with rules, people will focus on how to break them," she says.

Instead, she suggests starting with principles. "What are the values you want for commercial activity and international conduct in outer space? If governments and other stakeholders can agree on values, then they can discuss rules with a common framework of principles. The rules they develop will be more inherently acceptable to all and thus more likely to be followed."

This is, of course, a great deal easier said than done, but it's a roadmap that could lead to an effective outcome. The mechanics of it are open to suggestion at this point, but as of today, what might be optimal is for leaders from different silos to come together informally and develop statements of principles that can provide the basis for more formal talks—which, in turn, lead to agreements and the like. The process might take years, but it could start now.

Should Solutions Be Sui Generis or Adaptive?

As we have discussed throughout this book, space is similar to terrestrial domains, but different in distinctive ways. It's sort of like the sea, but not quite. It's sort of like Earth, but not quite. For this reason, the best approach may be to think about solutions to space crime, piracy, and governance that are sui generis, or original unto themselves, rather than adaptive of existing approaches on Earth.

We think there is room for both, certainly at this conceptual stage of solving a problem that hasn't yet come into existence. Why constrain ourselves to terrestrial legal frameworks if space is sufficiently different? The reality is that a similar progression of thought occurred on Earth over the preceding centuries. For example, there was a law of the sea in ancient times, but in that period, ships didn't travel very far. If you left ancient Rome for Greece, you were not more than day or two away from the nearest port. The law of the sea was suited to that environment.

As boats got bigger and better and could disappear across the globe for years at a time, the law of the sea had to be updated. The new laws of the sea, which developed to deal with situations like piracy in the Caribbean, which was far, far out of sight and mind to ship owners in Europe, were essentially sui generis. It was a new thing, even if it resembled the old thing. Space may be the same. Certainly, it's worth considering sui generis solutions for space piracy and crime. This is what we will be doing in each section of this chapter, where we consider possible ways to address the potential for space crime and piracy.

Accept the Limits of International Law, but Plan to Act Despite Them

The nations of the world come to the project of establishing laws in space with a significant disadvantage. Attempts to create something resembling the rule of law among nation-states on Earth have largely failed. At a minimum, their efficacy is inconsistent. Examples abound, but the recent ineffectiveness of international sanctions in dislodging Russia from its illegal annexation of Crimea stands out as a failure of the idea that international law will affect the outcomes of international conflict.

Looking at the potential for piracy in space, it would be wise, therefore, to accept that international law might have limited, or no, impact on the behaviors of bad actors. If those bad actors are in the employ (overt or stealth) of nation-states, then international law might influence them to some degree, but it would be foolish to assume that it will. Instead, the best course of action is to plan for the inapplicability of international law. What this will look like is hard to say, but it might feature kinetic countermeasures that punish wrongdoing regardless of international laws and norms.

Borrow and Improve upon Maritime Law and Norms

Maritime law offers a good starting point for thinking about policies and organizations that can establish and enforce laws in space. However, it would be a mistake to port maritime laws and norms to space uncritically. We could do better.

One basic problem with maritime law is that it's only partly a set of laws. It also encompasses a lot of traditions that just are...based on thousands of years of "this is how we've always done it." The legalities of salvage, for example, are rooted in ancient Rome's law of the sea.[1] The Romans were impressive legal minds, but there's an opportunity in space to retire 2,000-year-old legal precedents.

The flag of convenience is one maritime tradition that would be wise to avoid in space. This cat is already starting to sneak out of the bag, as we discussed in Chapter 6. It's not too late, however. As more spacecraft are produced and ready for registration, some strictures on flagging would be useful in curbing potential abuses and criminality in space.

Eliminating flags of convenience and correcting the sea's deficiency in law enforcement and courts would go a long way to preventing bad actors from evading regulations. There would be fewer "dark fleets" in space if such rules were in existence.

Treating space as a lawless area, akin to the open seas, would also be a mistake. This is easier said than done, for a variety of reasons, but if principles matter, this is a principle that deserves attention. Conduct in space should be subject to some agreed upon set of defined and enforceable laws.

This may all be moot in practical terms. As comparable efforts at coordinating international laws have shown, nation-states, corporations, and nongovernmental entities tend to do what they want to do and wait for regulation to catch up, if it ever does. That doesn't mean we should resign ourselves to failure, though. For one thing, there is a potentially strong financial case for creating strong rules for space commerce and settlement. That could create an incentive to cooperate.

Additionally, the current exclusivity of space travel creates a helpful bottleneck that favors regulations. Unlike the sea, on which virtually every nation on Earth can easily travel, space travel right now is limited to the United States, Russia, China, India, and Japan, give or take. With so few true players involved, it may be possible to establish laws and norms that benefit everyone. Latecomers would be under pressure to conform.

Build the Foundations of Space Governance

Confronting the threats of space crime and piracy means addressing the much bigger issue of space governance. It means grappling with the challenge of deciding what the governing authorities will be in space. It also means confronting the reality that there currently is no governing authority in space, at least not one with any real power.

Laws and policies intended to restrain criminal activity in space would therefore most likely be part of a broader framework of space sovereignty and law. And, to the extent that current frameworks are deficient, or non-existent, efforts to reduce the risk of space crime and piracy will also be deficient.

Luckily, we are not operating entirely in a vacuum. Organizations, legacy frameworks, and nascent agreements do exist, and they comprise a good starting point for creating a coherent, enforceable approach to space governance.

The most logical body to address these issues is the UNOOSA, particularly its Committee on Policy and Legal Affairs Section (CPLA). Their work is already underway, with resolutions and treaty updates under consideration at this time. While many deride the UN as ineffective and the organization certainly has its problems, UNOOSA does function as a forum for consensus on space matters. It may be the best option there is at this time.

UNOOSA, or whoever takes on these issues, however, will have their work cut out for them. One of the first orders of business should be to update the property rights and regulatory frameworks from the Outer Space Treaty, which are obsolete. That's not a knock on the treaty. It simply came into existence in an era when the kind of commercial exploitation of space we are engaging in now was hard to imagine.

As the paper "Governance in outer space: The case for a new global order," written by attorneys from the firm of Norton Rose Fulbright, notes, "There is currently an absence of a clear global space regulatory framework dealing with property and ownership rights, liability in the event of a collision, dispute resolution, and the licensing and registration of security interests."[2]

The Norton Fulbright lawyers make the important point that this "absence" is creating a vacuum of sorts, a situation where nation-states or ad hoc coalitions of countries will "go it alone," in their words, in realizing their distinct visions of space sovereignty. The two embryonic "go it alone" efforts in this regard are the Artemis Accords, signed by the United States, the United Kingdom, Canada, Japan, Australia, and other "Western" powers, and the nascent memorandum of understanding (MOU), signed by Russia and China for lunar exploration and commerce. This arrangement will inevitably lead to conflict.

As human history suggests, conflict is unavoidable, but it is easier to resolve conflicts if there is an agreed-upon set of principles, leading to a consensus on rules. UNOOSA, or whoever is tasked with this process, should understand the importance of bringing such rules into existence and garnering consensus on their validity. The first order of business, according to Norton Rose Fulbright, is to establish a uniform property rights framework for space.

What would this look like? A new international agreement on space sovereignty, regulation, and property rights might follow the existing Earth template of private ownership and national sovereignty within a commons. In the same way that a corporation like Exxon can own a gas well in Indonesia, a nation-state on islands in the global commons of the Pacific Ocean, so too could a mining company own a mine on an asteroid, with the mining operation treated as a U.S. sovereign territory within the commons of outer space.

Getting to success may seem like a heavy lift, but the good news is that there are some pretty strong incentives for countries to come together in space property rights. Everyone benefits, in economic terms, if there is a workable, enforceable framework. Then, if there can be consensus, followed by a coherent, enforceable framework for sovereignty and property rights, it should be possible to incorporate rules regarding criminal conduct in space.

That's the ideal, of course. Reality may turn out differently, if Earth's history is any guide. We can always aim high, but it seems improbable that space governance and law will be better than what we have on Earth at this time. And what we have on this planet is pretty deficient.

At best, preventing crime and resolving conflicts in space comes down to international law. As Richard M. Harrison and Peter Garretson point out in their book, *The Next Space Race: A Blueprint for American Primacy*, however, international law is not really law at all. Rather, it's a set of agreements and norms that countries follow if it's in their interest to do so. This can lead to a lot of disappointments, with countries signing and then ignoring treaties. Russia, for instance, signed treaties that theoretically outlawed its invasion of Ukraine. We're now heading into the third year of that bloody war.

International law can also result in spectacularly unexpected results. In 1914, for example, Britain honored its commitment to Belgian neutrality, established by the 1839 Treaty of London, declaring war on Germany and plunging the country into a conflict that resulted in 750,000 British soldiers dying and the beginning of the end of the British Empire. As Sir Edward Goschen, British Ambassador to Berlin, said at the time, "Just for a scrap of paper, Great Britain was going to make war on a kindred nation who desired nothing better than to be friends with her." That "scrap of paper" represented international law, and Britain considered it important enough to go to war over. Similarly surprising events may occur in space, too.

Establish Norms and Mechanisms for Orderly, Lawful Space Activity

It may turn out that practical norms and technical mechanisms will be more important than international agreements in ensuring orderly, lawful activity in space. It's one thing for long-winded conferences at the UN to forge agreements that countries may not follow. It's another to get an actual space vehicle into orbit safely and allow it do its business in a profitable fashion. Countering space crime and piracy will probably rely on the latter.

Build on Existing Laws and Regulatory Frameworks

The work of making space a more lawful and orderly environment for commerce is not starting with a blank slate. The United States, for example, already has certifications and licensing procedures in place through the FAA, FCC, and National Oceanic and Atmospheric Agency (NOAA) for space launches, space radio frequencies, and space remote sensing activities, respectively.

Efforts to expand or improve upon these existing regulations are underway, though they are not without controversy. In 2023, U.S. House of Representatives Space and Aeronautics Subcommittee Chairman Brian Babin (R-TX) and Chairman Frank Lucas (R-OK) introduced a bill called the Commercial Space Act of 2023. According to its sponsors, the legislation "updates and modernizes the government oversight of commercial space activities." The bill noted, "As this sector has been rapidly growing thanks to transformative innovation and significant investments in public-private partnerships, this legislation reduces regulatory red tape, promotes safety standards, and encourages technological advancements."

Chairman Babin said, "The country that leads the way in space exploration sets the standards for how we operate there. To continue strengthening American leadership in space, we must modernize our commercial space framework. The Commercial Space Act will do just that by addressing outdated laws, enhancing innovation, and ensuring our partners are not stymied by regulatory hurdles. The U.S. cannot afford to surrender ground in space by failing to act, which is why the passage of this bill is vital."[3]

Reaction to the bill has been mixed. Writing for JustSecurity.org, John Goehring, assistant general counsel at the U.S. Department of Defense (DoD) and a USAF Reserve Lt Colonel, expressed concern that the proposed law would have negative implications for national security. For example, he is worried that replacing the current NOAA certification of space-based sensing activities with a new, less regulated oversight by the U.S. Department of Commerce, would create an environment where it becomes easier to spy on American national security operations.[4]

On another front, NOAA announced contracts in 2024 to start creating a Traffic Coordination System for Space (TraCSS).[5] The idea is to move the responsibility of federal space situational awareness from the DoD to a civilian entity. It's early in what promises to be a multiyear program, but the initial contracts awarded are for technologies like a global radar network to collect and organize data on spacecraft and orbital debris, a network of optical telescopes, and software to forecast the locations of objects in low Earth orbit (LEO).

Establish a Space Commodities Exchange

Commerce in space, particularly transactions related to the extraction and refining of space raw materials, is potentially chaotic and risky. As of now, buyers and sellers of space products and services have no organized way to conduct business. This market failure is not good for the commercial development of space, and it makes the space industry vulnerable to crime and fraud. A space commodities exchange would help address many of these potential risks.

As envisioned and championed by Bruce Cahan of Stanford University and his nonprofit Urban Logic, a formalized commodities exchange would enable all market participants in the space economy to better visualize and predict market opportunities and risks by letting them trade standardized and reliable space commodities.

The "Space Commodities Futures Trading Exchange" conceptualized Cahan and described by him and his coauthors,[6] in an article in *New Space*, would adapt existing terrestrial market mechanisms to space. Thus, just as it is possible to buy a futures contract to purchase crude oil on the Intercontinental Exchange (ICE) at the "Brent" price, so too would it be possible to purchase a contract for a ton of lunar regolith—and have adapted legal and regulatory safeguards in place by trading on a regulated exchange established in the United States.

From a commercial perspective, a commodities exchange of this type would facilitate organized, knowledge-driven trading. According to Cahan, "Space industry will develop more rapidly and coherently when there are rules provided by an exchange. You get transparency, as in, 'Here's something you can buy from somebody and sell to somebody on a trusted basis'. You're not going to be lost in space without propulsion, so to speak."

A space commodities exchange would also reduce the potential for trafficking in stolen or sanctioned space cargoes or with untrustworthy parties. An exchange can't eliminate that risk, but it will make it harder for malicious actors to abscond with space commodities. The more cargoes that are committed to futures contracts, the fewer there will be to misappropriate.

According to Cahan, a space commodities exchange provides other market stabilizing benefits to its members: space debris can become a commodity recaptured, warehoused, repurposed, and resold among exchange members. Gaps in the International Space Treaties that today inhibit robust commercialization can be filled in and adapted among exchange members to provide greater certainty of their rights and remedies, the transparency of their commodities' origins and functionality, and other issues.

He went on to explain that risks that are today borne by space companies or their government or private-sector customers can be partially "offloaded" as derivatives—think swaps and insurance-linked securities—and resold on the exchange to parties who regularly trade in such risk-containing instruments. Finally, the exchange's transactions data provides greater certainty as to who is traded what space commodity for what orbit at what time, so as to make clear where and when supply and demand curves meet, which in turn justifies the levels and types of investment commercial space companies need to sustain their business plan and purpose.

Require Insurance Coverage for Acts of War and Piracy in Space

As we have learned over the last several thousand years, banning a practice does not stop it. Treaties and law can deter criminal behavior and provide a process for prosecution and punishment, but they don't stop crime. This has been amply demonstrated on the high seas. The risk of losing a vessel or cargo to piracy remains, despite the laws of the sea. Insurance fills this gap, covering ship owners and merchants from loss. The same could be the rule in space.

Insurance policies covering space cargoes for acts of war and piracy should be mandatory. Or, if not mandatory, it would be good if such insurance were baked into any space commerce transaction by custom. This is how maritime shipping works. It is possible to move goods on the sea without insurance, but in practice, it's almost never done, and for good reason. Space insurance policies, which now cover crashes, could easily be extended to cover piracy and acts of war.

Develop and Enforce Effective Antipiracy Cybersecurity Policies in Space

The vulnerability of space-based computer systems to cyberattack, described in Chapter 5, leaves many, if not all, space assets exposed to risk from hackers operating as space pirates. This problem is very serious, with the potential for Earth-based attackers to act asymmetrically and compromise expensive and strategic space assets with relatively little in the way of resources.

Space cyber vulnerabilities are comparable to the many cyber vulnerabilities found in Earth-based systems. And, as the slow, questionable progress in combatting such risks on Earth suggests, if we want to do better in space, we should start thinking quite differently about the issue now. Now is the time to map out a new, sui generis approach to space cybersecurity. It's clear, from the many failures in cyber defense on Earth, that applying an incremental set of fixes to space systems will result in disaster.

We are not alone in having this view. Writing in *Scientific American* in 2024, Sylvester Kaczmarek, CTO of OrbiSky Systems, said, "We need immediate, fortified satellite cybersecurity—not as a distant aspiration but as an urgent imperative right now. This is not a call for vague future planning but a demand for decisive action now to avert an all-too-likely scenario where critical services are incapacitated with far-reaching and devastating consequences."[7]

Harrison Caudill, founder of the Orbital Security Alliance, a collaboration between industry, academia, and government, to bring missing cybersecurity infrastructure to the space industry, had a similar perspective. In his paper "Big Risks in Small Satellites," he argued for a multipronged effort to make satellites more secure from cyberattack. In his view, success will come from a combination of factors, including specialization of trade and market incentives.[8]

Kaczmarek and Caudill, as well as ourselves, believe that cybersecurity in space will not be realized through today's open-ended, framework-based, and largely voluntary processes of establishing cybersecurity. No simple solutions are available, but it is clear what doesn't work well, so it would be wise to avoid the current approach, which involves developing a set of suggested standards and then auditing how well companies and government agencies follow them. That would be asking for failure.

Instead, the best approach would be a program that combines, as Caudill suggests, market incentives, with strict regulations. The aviation industry offers a model. For a computer system to run in the cockpit of an aircraft, it must first undergo strict certification processes. That would be a good idea in space; in other words, no computer goes into orbit if it isn't certified by a government agency of some kind. For commercial incentives, factors like insurance can be quite effective. For instance, if a space insurer will not underwrite a policy if a space asset lacks a certified computer system, that creates a strong motivation to install such a system.

Critics of this approach have said, in effect, well, good luck with that. Not everyone will follow the rules. That's true, but as with any security effort, processes and policies that reduce risk should be favored. If 80% of space-based computers are certified, versus zero, that's an achievement. It's better to deal with the 20% that are not well defended than to defend 100%.

Making an idea like this work is yet another "easier said than done" challenge, but our sense is that there is enough concern about the issue to bring government and private-sector entities together to solve the problem.

No one wants successful hacks and data breaches in space. Will that translate into action? We will have to see. The time to focus on it is now.

Visualize and Operationalize a Role for the U.S. Military and Other Armed Forces

Defending space commerce, transport, and settlement from pirates may be a job for the U.S. military, along with those of other countries. What that military role will look like, and how forces will be organized and led, remains to be seen. It's not an easy problem to address, as some current events reveal.

As of the writing of this book, the world is witnessing an assault on shipping in the Red Sea by Houthi rebels in Yemen. This small faction, backed by Iran, is wreaking havoc on global commerce, with billions of dollars' worth of oil and other cargoes rerouted around Africa's Cape of Good Hope versus risking a run at the Suez Canal.

Navies from the United States and United Kingdom have stepped up and unleashed a series of powerful air strikes on Houthi forces in Yemen. This has done little to stop the Houthis.[9] They continue to harass ships in the Red Sea. This disappointing outcome can tell us something about what it will take for the world's armed forces to combat space piracy.

If two of the world's most powerful navies cannot stop a small group of rebels from disrupting global commerce, that failure has implications for space. Dealing with pirates in space will be a more complicated mission than bombing the Houthis. The armed forces need to start thinking about this challenge today.

Solutions: The U.S. Space Force, Today and Tomorrow

The U.S. Space Force (USSF) is the only force in the U.S. military tasked with space-based threats. We can presume that the Space Force will be called upon to address space crime and piracy as it emerges in the future. As of today, this appears to present a challenge but also an opportunity.

The USSF, in its current configuration and mandate, is not set up to respond to the kind of criminal and pirate actions that are likely to occur. Their mission statement, "Secure our Nation's interests in, from, and to space," appears to cover such threats, in theory. However, the day-to-day activities of the Space Force are almost exclusively focused on managing and protecting satellites.

The USSF's current operational setup makes sense, given that it's focused on today's immediate responsibilities and risks. It may not be fair to judge such a new force for doing the limited job it has been assigned. That said, we believe there is room for a bigger picture, a broader vision of what the Space Force could become as novel threats arise in space.

It might be an interesting thought exercise to rethink the Space Force's mission with inspiration from the U.S. Navy's mission statement. That might look something like this: "The United States is a spacefaring nation, and the U.S. Space Force protects America in space. Alongside our allies and partners, we defend freedom, preserve economic prosperity, and keep space open and free."

To a certain extent, this is already happening with the U.S. Space Command (USSC), a unified combatant command whose personnel come from the USSF. According to its website, "The U.S. Space Command working with Allies and Partners, plans, executes, and integrates military space power into multi-domain global operations in order to deter aggression, defend national interests, and when necessary, defeat threats."[10]

The *New York Times* offered a useful way to differentiate the USSF from the USSC. The USSC, as their articled states, "reports to the U.S. secretary of defense." The USSC "is a decision-making entity that coordinates space-related operations for all branches of the armed forces. The Space Force, by contrast, outfits and trains troops for potential space conflicts, scrutinizes space for dangers and launches satellites for the DoD. The Space Force regularly consults with Space Command to ask how (or whether) the potential threats it discovers should be addressed."[11]

Between the USSF and the USSC, we have the makings of a military force capable of mitigating the threat of space crime and piracy. They're not quite there, though. The USSF is focused on satellites, and USSC appears to view space as a warfighting domain. That is of course necessary, but it may not be suitable for dealing with irregular, criminal-driven activities that affect American space assets.

There is plenty of time to think about how USSF and USSC might approach space crime and piracy. For insights, we turned to Tyler Bates, a former U.S. Air Force and Space Force officer who has analyzed irregular warfare in a personal capacity. He offered some thoughts on what a revised mission for the USSF and USSC might look like in this regard.

When it comes to mitigations, Bates' view is that:

- The USSF needs capabilities and expertise to counter space crime and piracy challenges.
- Conventional USSF forces need various means to defend themselves and others from predation and other aggressive actions in space. These include space domain awareness (SDA); lunar intelligence, surveillance and reconnaissance (ISR); space mobility; competencies and capabilities for robotics; AI; and cyber and information operations for space.
- The USSF also needs special operations forces (SOF) and the ability to conduct its own irregular warfare operations to counter those of an adversary or competitor.
- The USSF and USSC should form a Combined Task Force for countering piracy in space alongside allied and partner nations (i.e. the space version of the sea-based combined task force CTF-151.)
- The USSF also needs law enforcement authorities, capabilities, and expertise. This could be possible based on private American ownership of space assets based on the Artemis Accords, which could confer law enforcement authority on the USSF. (The Posse Comitatus Act is not applicable in space, so there are no strictures on a branch of the U.S. military enforcing the law in space.)
- The USSF and USSC also need to partner with U.S. Marshals to hunt/ transport fugitives in space and sell forfeited/detained assets. U.S. Marshals could potentially also deputize people to act as U.S. deputy marshals within U.S. spacecrafts, space stations, or planetary surface facilities or vehicles. (An analogous partnership occurred in Antarctica, a territory on Earth with similar legal characteristics to space.)

Establish a U.S. Space Guard

An alternative to commissioning USSF and USSC to combat space piracy is to develop a "Space Guard" to perform this role. This idea has been championed by Anna Gunn-Golkin, a USSF officer now serving as a White House fellow. Her 2018 article, "Space Guardians," published in *The Space Review*, outlines a vision for establishing a U.S. Space Guard (USSG) with a mission similar to that of the U.S. Coast Guard but applied to the space domain.[12]

As she explained in her article, the USSG would be a law enforcement agency that will be "ready to respond and enforce the laws when needed" and oversee the licensing of American spacecraft. The USSG would also come to the aid of ailing spacecraft, as the Coast Guard does today. It will "perform space security operations, protecting U.S. astronauts, ships, and cargo." The advantage of a hypothetical USSG is that its mandate for antipiracy activities would be clear. The Guard would be organized for it and structure its resources and chain of command to operationalize the mission.

One issue with this idea, however, is that the laws she envisions being enforced by the USSG do not quite exist yet. As we have discussed, laws governing conduct in space are subjective and full of gaps that criminals will exploit. A Coast Guard cutter can interdict a smuggler because that smuggler is violating a known law. The space version of that scenario is a lot less clear.

Visualize and Operationalize a Role for the Intelligence Community

Detecting, preventing, and responding to acts of space piracy that could affect national security is a job for the nation's intelligence community (IC). This is a secretive world, so it's difficult to say with any certainty how well they are doing with this possible assignment. (Indeed, the USSF has the highest proportion of classified budget of all U.S. military forces.)[13] However, based on conversations with insiders familiar with these matters and observations we can make based on publicly available information, it seems the IC could be more optimally engaged in a productive way on this important issue.

The stakes are potentially high, as the investigation into Israel's major intelligence failure in the wake of the deadly Hamas attack on October 7, 2023, reveals. According to reporting in the *Jerusalem Post*, the probe into what went wrong in Israel's famed Unit 8200 is split between blaming the unit's commander, known as "Y," and examining underlying structural factors in the organization.

As the article says, "The other faction offered a different, opposite perspective, cautioning against a rush to judgment. They argued that 'to single out 'Y' for blame is to misunderstand the nature of the failure we're confronting. It's not just a question of individual error but a systemic flaw, a problem rooted in decisions made long before 'Y' took command'. They insisted that accountability should extend beyond and that systemic changes are needed to prevent future failures."[14]

Space intelligence is where Unit 8200 was a decade ago. Decisions made today will cast long shadows. Will "systemic flaws" lead to ineffective space intelligence? That is the question we need to be asking today, even if the adversaries and threat scenarios being contemplated are years off in the future.

With that in mind, what follows are some thoughts and suggestions for making intelligence a valuable contributor to positive outcomes regarding space piracy. This means looking at space intelligence more broadly. Combating crime, piracy, and smuggling in space should be part of a much bigger picture of engaging in effective intelligence gathering and operations in space. We look at three areas of concern: overall intelligence strategy, intelligence gathering and sharing, and operations. All three are necessary if the IC is to play a useful role in dealing with the threat of space piracy.

Space Piracy, Intelligence Gathering, and Overall Intelligence Strategy

Space is a major priority for the IC. The National Reconnaissance Office (NRO), for example, collects data on America's adversaries, working closely with the CIA and other intelligence agencies. The CIA gathers intelligence about other nations' space programs.[15] In 2021, the USSF's Intelligence, Surveillance, and Reconnaissance (ISR) Enterprise was formally added to the U.S. IC, joining the DIA, DEA, FBI, and others to become the eighteenth IC agency.[16] The USSF also has its Space Operations Command (SpOC), which operates spy satellites and processes satellite imagery in support of other branches of the U.S. military.[17]

As a member of the IC, the USSF provides early warnings of threats to U.S. interests, assessments of adversary space technology roadmaps, and insight into adversary intentions in the space domain. The force focuses on the collection and analysis of adversary space and counterspace technologies, and assessments of adversary operations enables a decision advantage for leadership at all levels.

As a member of the Joint Force and IC, the USSF routinely coordinates with our sister services in the DoD as well as organizations like the NRO. In practical terms, the USSF prepares to respond to a range of threats in space from nonkinetic and reversible effects like jamming to kinetic operations like antisatellite missiles.

The question to ask, in our view, is whether this current setup is optimal for handling the full spectrum of space-related threats. As of now, based on our conversations with people who are familiar with space's current role in the IC, it appears that the focus is almost entirely on major sovereign adversaries like China and Russia.

This seems limited, and it would be advantageous to expand the portfolio to include secondary—but highly relevant—sovereign actors like Japan, Germany, India, Iran, Israel, and others. From there, the portfolio could expand to include intelligence concerning nonstate actors such as criminal groups, terror groups, and so forth. We're not saying the IC isn't paying attention to these entities. We're saying perhaps they could balance their priorities and do more.

If the IC is to pursue such a broad mission, a follow-on question might then be: "Where's the IC's center of gravity?" Who will take the lead? We think it should be the USSF. In the same way that the U.S. Navy is the center of gravity for naval intelligence, so too can the USSF be the focal point and locus of control for all intelligence activities related to space.

The USSF as it is now constituted does not appear suited to this mission. Part of the problem is structural. Security in space has a lot to do with the security on Earth. We launch rockets from land- and sea-based platforms, for instance. These are targets for space pirates and other bad actors, so the USSF would need to develop deep intelligence capabilities to ferret out threats. Given the United States' overall strategic goal of achieving supremacy in all fighting domains, it is worth asking if the USSF's operational mandate be expanded to include Earth, for example, cultivating human intelligence (HUMINT) assets on Earth who can speak to threats from space.

Such a strategy would also necessitate strong collaboration and interoperation with other branches of the military. These exist today, of course, but if the USSF is to be the "center of gravity" for space intelligence, including relevant Earth-based intelligence, these bonds will have to become stronger and tighter.

Success might also mean having the USSF "borrow" some strategic and operational intelligence ideas from the U.S. Navy and other service branches. For example, the Navy, through its Office of Naval Intelligence, understands crime and piracy because fighting pirates is part of the Navy's overall mission.

Getting to success for the USSF as the center of gravity for space intelligence might also involve rethinking certain aspects of its force structure and operational programs. Most of the USSF's current activities are technology-focused, with Guardians working at desks operating computers. This is necessary for their workloads, but it is not ideal for developing the kind of "warrior spirit" that a service branch needs to become a passionate defender of the United States.

This is not a criticism or an accusation. Far from it. We are simply pointing out what we and others perceive as a deficiency, borne of the service's origins, that could be remediated to become a more aggressive player in the intelligence and warfighting arenas. Further to this point, we think the USSF could learn from the CIA and its combination of intelligence gathering, covert, and clandestine operations.

Space Intelligence Operations

What will it take for the IC to be effective in intelligence operations in space? This is a question that requires answers that are strategic but also pragmatic. Returning to our dialogue with "Bob," a former operative in the IC, he highlighted how actions like threat response require clear definition of authority and accountability—but also specific capabilities.

Bob's long career in intelligence has given him a useful perspective on how the IC can be focused to respond to novel threats. He's lived through the IC's adaptation to airline hijacking, for instance, which caught the IC off guard in the 1970s. Space will be a similar evolution, in his view.

He tried to help us visualize how the IC and the government in general would respond to an act of space piracy. "Typically, when something like this that is totally outside the box occurs," he said, "the first thing that happens is that there is denial. Like, 'It didn't happen. Somebody's making up a report, etc.' Eventually, as information comes in, you go from denial to a bit of anger, rather displaced anger—anger at the fact that we didn't get out, that we don't have the facilities to deal with it, we don't have the laws."

He elaborated, explaining that as of now, as far as he knows, the United States has no capability to deal with an act of space piracy. The FBI has a hostage rescue team. They're very good at what they do, but they're trained to work here on Earth. They couldn't occupy anything in space. The FBI also has a section that creates psychological profiles of hostage takers, but they're not set up to do that for space pirates…yet.

The CIA has a covert action staff, but they would be unable to get to space. "And even if they got there, they're not trained to operate in space. No one is," he added. The President has Seal Team Six at his direct command. "They, and other elite special operations groups are at the President's beck and call, but they too, are not trained for space, and they can't get there. As a result, there'll be a crime in space that no one can handle."

Dealing with space piracy will require a paradigm shift in the IC, according to Bob. And that, as he explained, will take some effort to execute. There are precedents, however. He compared the potential space piracy threat to the advent of airline hijacking in the 1970s. "Just as air piracy led to the creation of the TSA, with space piracy, we'll have to bring on a brand-new agency of the government that deals in space crimes."

Under the law, such a new agency would almost certainly be part of the USSF, but it would probably operate semiautonomously. It might need to operate outside of official channels. It would need a very broad brief, able to engage in data collection and clandestine operations across a wide range of terrestrial and space domains. Operatives might cultivate intelligence sources in rocket propulsion companies, for example, or inside venture capital funds. These are the sort of unlikely places where large-scale acts of space crime could take shape.

At this point, we are simply sharing ideas that may or may not hold up to serious scrutiny. However, we believe that it's essential to have these dialogues now. That way, the process can reveal the best approach and the hard work of making it happen can begin.

Takeaways

This chapter attempts to answer the most basic and necessary question that arises from an analysis of space piracy: what should be done about it?

Assuming that the problem is real, it seems clear that current treaties, frameworks, and institutions are not set up to deal with preventing and responding to space crime and piracy.

That said, it's not as if the elements required to combat space piracy are nonexistent. The Outer Space Treaty and other agreements provide a foundation for more detailed future agreements. The Law of the Sea provides precedents that might be useful in space, though there is the potential to do better than maritime law's clumsy traditions. International law in general is weak, but it is better than nothing. The best approach is to leverage what we have and move ahead by bringing stakeholders together to formulate a variety of laws and practices that will mitigate the risk of space crime and piracy.

Who does this work involve? It's a multithreaded issue. Corporations, governments, law enforcement bodies, diplomatic entities, insurance and finance businesses, militaries, and intelligence services should all have a say.

How this would work remains to be seen. The UN is probably the best venue for the basic work, though other bodies may need to come into existence to realize the principles and rules that stakeholders develop. Finding consensus on space governance is a key gating factor here. In other words, how do space-based entities govern themselves? Who is in charge? What are the laws? All of these questions need to be answered, among many others.

The process results in a range of "easier said than done" tasks, but they need to be pursued, even if the results are deficient. Practicalities and norms also matter, perhaps more than high-minded policies. Establishing a space commodities exchange, for example, could bring meaningful order to space commerce and reduce the risk of piracy. Requiring insurance and cybersecurity certifications would also help.

The U.S. military and IC also need to play a role in combatting space piracy. They would be the "teeth" of any antipiracy policies and practices. While it's likely that private security firms might fill in the gaps, there's no substitute for the real thing. The question is whether they will consider the issue serious enough to focus on at this point. This remains to be seen. Also, there's a lot that's unknown about what the USSF is doing in this regard.

Some observers believe the force is not well set up for combatting space piracy at this time, given their current mandate. There is time to figure these challenges out, however. The threat is most likely coming, but if we take action now, we will be prepared for it.

Notes

1. Emilia M. Ferrándiz, "A Sea of Law: The Romans and their Maritime World," American Society of Overseas Research 11, no. 5 (May 2023), `https://www.asor.org/anetoday/2023/05/sea-of-law`.
2. Scott Atkins, Martyn Taylor, Holly McAdam, Richard Morrison, and Jo Feldman, "Governance in Outer Space: The Case for a New Global Order," Norton Rose Fulbright, November 2022, `https://www.nortonroseful` `bright.com/en/knowledge/publications/e8862684/` `governance-in-outer-space-the-case-for-a-new-global-` `order`.
3. Babin and Lucas Introduce Legislation to Modernize Commercial Space Sector, Before the Committee on Science Space and Technology, 118th cong. (November 2, 2023), `https://science.house.gov/2023/11/babin-and-lucas-` `introduce-legislation-to-modernize-commercial-space-` `sector#:~:text=H.R.,oversight%20of%20commercial%20` `space%20activities`.
4. John Goehring, "The Commercial Space Act of 2023 is Bad for National Security," Just Security, December 19, 2023, `https://www.justsecurity` `.org/90567/the-commercial-space-act-of-2023-is-bad-` `for-national-security`.
5. Tim Fernholz, "Commerce Takes the First Step Towards Civil Space Traffic Coordination," Payload, January 22, 2024, `https://payloadspace.com/` `commerce-takes-the-first-step-towards-civil-space-traffic-` `coordination/?oly_enc_id=9807J1446578A7S`.
6. Cahan, Pittman, Cooper, and Cumbers, "Space Commodities," 211.
7. Sylvester Kaczmarek, "We Need Cybersecurity in Space to Protect Satellites," Scientific American, February 5, 2024, `https://www.scientificamerican` `.com/article/we-need-cybersecurity-in-space-to-` `protect-satellites`.
8. Harrison Caudill, "Big Risks in Small Satellites: The Need for Secure Infrastructure as a Service" Conference paper, ASCEND 2020: Cybersecurity Practices, Lessons Learned and Case Studies, November 16–18, 2020. `https://doi` `.org/10.2514/6.2020-4017`.

9. Joshua Cheetham, Shruti Menon, Yi Ma, and Paul Myers, "US and UK Strikes Fail to Slow Houthi Attacks," BBC, February 1, 2024, https://www.bbc.com/news/world-middle-east-68159939.

10. "Home: U.S. Space Command," United States Government, accessed August 10, 2024, https://www.spacecom.mil.

11. Jon Gertner, "What Does the U.S. Space Force Actually Do?," The New York Times Magazine, November 8, 2023, https://www.nytimes.com/2023/11/08/magazine/space-force.html.

12. Anna Gunn-Golkin, "Space Guardians," The Space Review, June 25, 2018, https://www.thespacereview.com/article/3520/1.

13. Gertner, "U.S. Space Force."

14. Ben Caspit, "Inside Unit 8200: Moving Forward After the October 7 Intelligence Failure," The Jerusalem Post, February 25, 2024, https://www.jpost.com/israel-news/article-788828.

15. Brett Tingley, "The CIA Knows a Lot about other Nations' Space Programs. You can too with its New 'World Factbook' Update," Space, August 29, 2023, https://www.space.com/cia-factbook-world-space-programs.

16. Loren Blinde, "U.S. Space Force becomes 18th IC Agency," Intelligence Community News, January 11, 2021, https://intelligencecommunitynews.com/u-s-space-force-becomes-18th-ic-agency.

17. Alexandra Lohr, "Space Force Moves to Consolidate its Intelligence Community," Federal News Network, October 24, 2022, https://federalnewsnetwork.com/space-operations/2022/10/space-force-moves-to-consolidate-its-intelligence-community.

Conclusion: We Need to Talk

European statesmen meeting at the Congress of Vienna to close the Napoleonic Wars, 1815. Hand-colored woodcut.
Source: Licensed from alamy.com

Writing this book has been an experience akin to attending a fascinating symposium where we got to speak to a lot of very smart and knowledgeable people. We shared our ideas and heard theirs. We've tried to capture the essence of this integrative, multithreaded dialogue in these pages. We hope we've succeeded.

Like many authors, we are concluding the writing process with different thinking about the issues than we had at the start. That's a good thing. We've learned something in the process.

That said, our basic convictions regarding space piracy remain unchanged. To us, it's a serious threat, one that requires a significant commitment from multiple parties to mitigate. We hope this book will catalyze the conversations that need to take place for this purpose. Ideally, if you've read this far and have a role to play in combatting space piracy, you're thinking "We need to talk…"

Recapping the Current Situation and the Risks of Space Piracy

The paradox of this book is that we've managed to get worked up about a problem that doesn't exist yet. Several well-meaning colleagues have made this gentle suggestion to us: Why are you so agitated about this? We may not face space piracy for decades, if ever. Yet, space piracy is emerging, currently in the form of hacking. The future seems almost guaranteed to deliver more potent forms of space-based criminality.

Potential space piracy scenarios include satcom disruptions and interference in aviation, broadcasting, and telecom. We could start to see kinetic attacks on Earth-based space assets, such as launch sites. We may soon start to experience hijacking and ransoming of satellites, and more.

When these events start to occur, no one who is familiar with the history of piracy will be surprised. For millennia, valuable cargoes transiting narrow passages have been the targets of pirates. From Biblical times to the Roman Empire and recently at the Horn of Africa, pirates flourish where trade exposes pricy assets to thievery.

Piracy's "Golden Age" in the seventeenth and eighteenth century reveals how piracy can play a role in wars between great powers—as well as piracy's integration into national financial and monetary systems. Space today is comparable to the Caribbean in that era. It's remote and lawless, open to predation.

Will space become home to enough valuable assets to attract criminals? The growth of space industries strongly suggests this to be true, though the timing and details of this growth remain unclear. We are seeing large investments into new satellite ventures, as well as funding for new business concepts like satellite servicing, in-space manufacturing, space mining, and new propulsion technologies. Even far-fetched ideas like space-based energy generation are getting serious consideration for investment. As these industries mature and become profitable, they will be targets for pirates.

Space-based computer systems, and computers on the ground that support them, are vulnerable to cyberattacks. This may be where space piracy begins. The risks are here today, driven by the many cyber vulnerabilities affecting all computer systems and networks. Any effort to combat space piracy should concern itself deeply with space cyber risk. Cybersecurity also offers a cautionary tale about the unexpected consequences of engineering decisions. The world is in a state of cyber insecurity, one that threatens our entire society and costs hundreds of billions to fix, due to mistakes made 60 years ago. We can avoid repeating this blunder in space if we focus on the issue now.

As far as law in space is concerned, the most generous opinion we can offer is that it's vague and incomplete. While space is not a lawless environment, technically speaking, it is in fact an environment where Earth-based laws are difficult to define and essentially impossible to enforce. The Outer Space Treaty and other agreements provide a basic framework for space law, as it might relate to piracy, but a lot of work needs to be done in this area. Risks abound, including "flags of convenience" and jurisdictional gaps.

Space pirates may come from the ranks of criminal cartels, especially if space piracy occurs on the ground. These groups have the resources to execute space piracy, if they so choose. Not everyone agrees with this prediction, with some experts saying that criminal cartels are more interested in present, easy opportunities versus technically complex ventures like space crime.

Space piracy, if it comes to pass, will affect the national security of the United States and other countries. Given the criticality of satellites to military operations and any number of other essential industries, any threat to satellites is a threat to the country's security. Indeed, space piracy may be a preferred method of geopolitical aggression because, like the hacking we're seeing against critical infrastructure today by Russia and China, it provides deniability.

The commercial sector is also at risk from space piracy. Crime in space has the potential to disrupt the insurance, banking, and finance industries. Space may be a good venue for money laundering, for example, with impacts on banking regulations. Corporate activities in space could also lead to space piracy, such as through the evolution of private security firms and mercenaries into space powers unto themselves. Some of this is speculative and futuristic, but the scenarios are worth exploring as they do represent true risk to commercial endeavors in space.

Ultimately, we think it's safe to draw the conclusion that the threat of space crime and piracy is real. And we can further assert that existing treaties, frameworks, and institutions are not ready for the challenge. There is much work to do.

Conclusions and Calls to Action

What needs to happen if we are to be prepared for space piracy when it becomes a reality in the coming years? And who should be doing the preparing? As we have said, no single entity or constituency of stakeholders can go this alone. It will have to be a group effort.

The work of preparing for space piracy will fall to industries that operate in space or adjacent to it, as well as government institutions, law enforcement agencies, diplomats, insurance and financial institutions, and the military and intelligence community. This is a big group, and the list is likely incomplete. But it's real. As we have seen in dealing with piracy off the Horn of Africa, a lot of different entities have to partner up for success.

From there, it's about building on existing treaties, laws, and practices. We can borrow from precedent, such as the Law of the Sea, though with a mind to getting it right this time around. We can lean into international law, but accept its weaknesses and try to build something better for space. This may sound naïve, but given the potential for profit in space, there are incentives for countries to cooperate. Conversely, the costs of disasters in space are so high that they, too, create pressure to develop a workable legal framework to prevent and prosecute crime in space. This can be done.

The locus of control for such an effort is unknown, but the United Nations (UN), its deficiencies aside, is probably the best place to start. As we said earlier, the UN is likely to be at the center of developing a range of agreements that govern conduct in space, set the parameters for sovereignty, and so forth. Dealing with crime, smuggling, and piracy in space could fit within these larger efforts.

A lot of this comes down to governance. Who will be in charge in outer space? Whose laws will be in effect? What principles will guide human activities in space? These questions are essential to the success of the space industry and space exploration in general, but they are also key to figuring out how to mitigate the threat of space piracy.

The U.S. military and intelligence community also have parts to play in this process. It would be wise for the relevant leaders in these arenas to contemplate the need for "irregular warfare" capabilities and the ability to respond kinetically and rapidly to acts of piracy. Services branches could learn from one another. The Navy has a lot to teach the Space Force about pirates. The Army could help strategize and prepare for the Earth-bound dimensions of space piracy.

The practical is also relevant. A space commodities exchange could bring stability and regulation to an otherwise chaotic commercial environment. So could clear banking rules for space. Insurance could enable commercial actors in space to be confident that their losses could be recovered if they encounter pirates, and so forth.

The main thing, though, is to get the dialogues started. We need to talk. Anyone who has a vested interest in space commerce, space exploration, or space colonization should consider joining the critical dialogues that need to take place about the piracy threat.

And when? Now! Now is the time to start the thought process and conversations that will lead us to meaningful results. The process could take years, but we have years. There is no rush, but events could gain on us, and we could find ourselves in reactive mode—dealing with problems that we could have brainstormed solutions for but didn't. Let's not do that. Let's get to work.

Bibliography

Ayres, Thomas. "A Maritime Solution for Cyber Piracy." *The Wall Street Journal*, May 13, 2021. https://www.wsj.com/articles/a-maritime-solution-for-cyber-piracy-11620922458.

Bloch, Marc. *Strange Defeat*. W.W. Norton & Company, 1968.

Bowen, Bleddyn. *War in Space*. Edinburgh University Press, 2020.

Brooks, Chuck. "The Urgency to Secure Cyber Secure Space Assets." *Forbes*, February 27, 2022.

Brown, Stephen R. *Merchant Kings*. Thomas Dunn Books, 2009.

Butcher, William. *Jules Verne*. Thunder Mouth Press, 2006.

Carlson, Joshua. "Spacepower Ascendant: Space Development Theory and a New Space Strategy." Independently published, 2020.

Chet, Guy. "Maritime Law as propaganda: The Case of Piracy Suppression in the British Atlantic." *World History Connected* 19 no. 1 (2022).

Chet, Guy. "Persistence of Piracy." Posted June 16, 2021. YouTube, 38:45. https://www.youtube.com/watch?v=69-MlDrB3Gg.

Coll, Steve. *Private Empire: Exxon Mobile and American Power*. Penguin Books, 2013.

Correa-Cabera, Guadalupe Los Zetas. *Criminal Corporations, Energy and Civil War in Mexico*. University of Texas Press, 2017.

Davis, Hank, and Christopher Ruocchio (eds). *Cosmic Corsairs*. Baen, 2020.

Emery, Andrew J. (Major, USAF). "The Case for Spacecrime: The Rise of Crime and Piracy in the Space Domain." Thesis, Air University, 2013. https://apps.dtic.mil/sti/tr/pdf/ADA615000.pdf.

Frankel, Earnst. *The Dual State*. Oxford University Press, 1941.

France, Martin E.B., and Jerry Jon Sellers. "Real Constraints on Spacepower." In *Toward a Theory of Spacepower*, edited by Charles D. Lutes and Peter L. Hays. National Defense University Press, 2011. https://ndupress.ndu.edu/Portals/68/Documents/Books/spacepower.pdf.

Franck, McKenzie. "Falling Stars and Sinking Ships: How Maritime Law Fills the Gaps of the Outer Space Treaty." *Pace International Law Review*. April 11, 2022.

https://pilr.blogs.pace.edu/2022/04/11/falling-stars-and-sinking-ships-how-maritime-law-fills-the-gaps-of-the-outer-space-treaty.

Fischer, William B. *The Empire Strikes Out: Kurd Lasswitz, Hans Dominik and the Development of German Science Fiction*. Bowling Green University Popular Press, 1984.

Garittee, Jerome. *The Republics' Private Navy*. Wesleyan University Press, 1967.

Giess, Robin, and Anna Petrig. *Piracy and Armed Robbery at Sea*. Oxford University Press, 2011.

Goguichvili, Sophie, Alan Linenberger, and Amber Gillette. "The Global Legal Landscape: Who Writes the Rules on the Final Frontier?" *Wilson Center*, October 2021. https://www.wilsoncenter.org/article/global-legal-landscape-space-who-writes-rules-final-frontier.

Goswami, Namrata, and Peter Garretson. *Scramble for the Skies: The Great Power Competition to Control the Resources of Outer Space*. Lexington Books, 2022.

Grant, Catherine L., James A. Russell, and Alessio Patalano (eds.). *The New Age of Naval Power in the Indo Pacific: Strategy, Order, and Regional Security*. Georgetown University Press, 2023.

Gunn-Golkin, Anna E. (Major, USAF). "Countering Space Pirates in the United States Space Guard." Air University, 2018.

Hanna, Mark. *Pirate Nests and the Rise of the British Empire, 1570–1740*. University of North Carolina Press, 2015.

Harris, Mark. "FCC Accuses Stealthy Startup of Launching Rogue Satellites." *IEEE Spectrum*, March 9, 2018. https://spectrum.ieee.org/fcc-accuses-stealthy-startup-of-launching-rogue-satellites.

Johnson-Freese, Joan. *Space Warfare in the 21st Century*. Routledge, 2016.

Klein, John J., *Understanding Space Strategy*. Routledge, 2019.

Konstam, Angus. *Pirates*. Lyons Press, 2008.

Konstam, Angus. *Scourge of the Seas*. Osprey Publishing, 2007.

Kraska, James. *Contemporary Maritime Piracy*. Prager, 2011.

Libicki, Martin C. *Cyberspace in Peace and War, Second Edition*. Naval Institute Press, 2021.

Lospinoso, Josh. "Space Race Needs Better Cyber Security." *The Hill*, January 13, 2022. https://thehill.com/opinion/cybersecurity/589542-space-race-needs-better-cybersecurity.

Macintyre, Donald, *The Privateers*. Raly Elek, 1975.

Mahan, Alfred Thayer. *The Influence of Sea Power Upon History, 1680–1783*. Digireads.com Publishing, 2020.

Mallett, Michael. *Mercenaries and Their Masters: Warfare in Renaissance Italy*. Pen & Sword Military, 2009.

McFate, Sean. *The New Rules of War*. William Morrow Paperbacks, 2019.

MacMillan, Ken. *Sovereignty and Possession in the English New World: The Legal Foundations of Empire, 1576–1640*. Cambridge University Press, 2009.

McNight, Terry, RFADM, USN (Ret.), and Michael Hirsh. *Pirate Alley: Commanding Task Force 151 Off Somalia*. US Naval Institute, 2017.

Ritchie, Robert. *Captain Kidd and the War Against Pirates*. Harvard University Press, 1967.

Ritter, Michel, and William MacLachlan. "Piracy." *The Shipping Law Review, Ninth Edition*, 2022.

Rotella, Sebastian, and Kristen Berg. "How a Chinese Gangster Transformed Money Laundering for Drug Cartels." *ProPublica*, October 11, 2011.

Sakhuja, Vijay. "Maritime Order and Piracy. "*Strategic Analysis* 24 no. 5 pages 923–938. 2000. https://doi.org/10.1080/09700160008455259.

Sankey, Margaret. *Blood Money: How Criminals, Militias, Rebels and Warlords Finance Violence*. Naval Institute Press, 2022.

Sivolella, Davide. *Space Mining and Manufacturing: Off-World Resources and Revolutionary Engineering Techniques*. Springer Praxis Books, 2019.

Szymanski, Paul S. *The Battle Beyond: Fighting and Winning the Coming War in Space*. Amplify Publishing, 2024.

Thomson, Janice. *Mercenaries, Pirates, & Sovereigns*. Princeton University Press, 1994.

Toll, Ian W. *Six Frigates*. W.W. Norton & Company, 2006.

Trease, Geoffrey. *The Condottieri*. Holt, Rinehart and Winston, 1971.

Triptree, James. *Her Smoke Rose Up Forever*. Taychon Publications, 2004.

Triptree, James. *Meet Me at Infinity*. Tom Doherty & Associates, 2000.

Tseng, Po-Hsing, Zhao-Chao Her, Nick Pilcher. "Piracy Defense Strategies for Shipping Companies and Ship: A Mixed Empirical Approach." *Maritime Transport Research* volume 2. 2021. https://www.sciencedirect.com/science/article/pii/S2666822X21000125.

Ward, Ralph. *Pirates in History*. York Press, 1974.

Weeden, Brian, and Victoria Samson (eds). *Global Counterspace Capabilities: An Open Source Assessment*. Secure World Foundation, 2018. https://swfound.org/media/206118/swf_global_counterspace_april2018.pdf.

Whisker, James, and John Coe. *Piracy Ancient and Modern*. Independently published, 2020.

Zegat, Amy. *Spies Lies, and Algorithms*. Princeton University Press, 2022.

Zubrin, Robert. *The Case for Space: How the Revolution in Spaceflight Opens Up a Future of Limitless Possibility*. Prometheus, 2019.

Acknowledgments

A book like this, which draws on knowledge from so many different fields, is truly a team effort. The authors wish to thank many individuals for contributing their time and deep expertise to this project. Our team at Wiley, Jim Minatel, Tom Dinse, and Sara Deichman, have provided wonderful guidance and direction for this project. Gordon Roesler has been one of our key advisors from day one, encouraging us and providing us with a perspective that we could not have gained elsewhere. Our friend "Bob" also guided us in understanding the opaque world of intelligence. We also want to acknowledge contributions from George Sowers, Denis Bensoussan, Johathan Gutoff, Salvatore Mercogliano, Guadalupe Correa-Cabrera, Dennis Silin, Christopher Newman, Frans von der Dunk, James Cambias, Toni Weisskopf, Michelle Bockmann, Phil Stern, Mark Hanna, Patrick Lin, Bruce Cahan, Martin Libicki, George Khouri, Mathieu Bailly, Jim Plaxco, Kenneth Rijok, Elinor Garcia, Forrest Meyen, Annie Pforzheimer, Naeem Altaf, Margaret Sankey, Nathan Jones, George Long, Tudor Stompf, Timothy Nelson, Tyler Bates, Aaron Moore, Loretta Napoleoni, Peter Garretson, and Peter Cook. We could not have done this without your help, and for this, we are very grateful.

About the Authors

Marc Feldman is executive director of the Center for the Study of Space Crime, Piracy, and Governance (CSCPG). For the last ten years, Marc has been involved in aerospace/space and defense-based venture capital. He has more than thirty-five years of experience in commercializing technologies and scaling startups. Marc has led teams across various industries, including life sciences, entertainment, media, telecommunications, consumer products, and aerospace/space. Having worked at and advised Shamrock Capital, Disney, Interpublic Group, Univisa Satellite Services, and News Corp, Marc has extensive experience globally, including in Asia, Europe, Russia, the Middle East, and Latin America.

Hugh Taylor is the author of 10 books about cybersecurity, business, and technology. A Certified Information Security Manager (CISM), Taylor was executive editor of the *Journal of Cyber Policy* for six years. As a writer, Hugh has created content for such clients as Microsoft, IBM, SAP, HPE, Oracle, Google, and Advanced Micro Devices. He previously served in executive roles at Microsoft, IBM, and several venture-backed technology startups. Hugh was a lecturer at UC Berkeley's School of Information and School of Law from 2007 to 2011.

Index